BURIED DREAMS

BURIED DREAMS

ANDREW R. BLACK

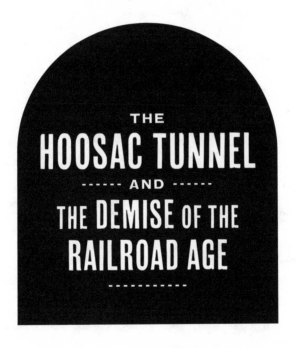

THE
HOOSAC TUNNEL
······ AND ······
THE DEMISE OF THE
RAILROAD AGE

LOUISIANA STATE UNIVERSITY PRESS

BATON ROUGE

Published with the assistance of the V. Ray Cardozier Fund

Published by Louisiana State University Press
www.lsupress.org

Manufactured in the United States of America
First printing

DESIGNER: Michelle A. Neustrom
TYPEFACE: Whitman
PRINTER AND BINDER: Sheridan Books, Inc.

Cover image: Detail of *View of North Adams, Berkshire County, MA, 1881,*
printed in 1881 by H. H. Rowley & Co. *Library of Congress.*

LIBRARY OF CONGRESS CATALOGING-IN-PUBLICATION DATA

Names: Black, Andrew R., 1943– author.
Title: Buried dreams : the Hoosac Tunnel and the demise of the railroad age /
Andrew R. Black.
Description: Baton Rouge : Louisiana State University Press, 2021. | Includes bibli-
ographical references and index.
Identifiers: LCCN 2019055134 (print) | LCCN 2019055135 (ebook) | ISBN 978-0-
8071-7357-2 (cloth) | ISBN 978-0-8071-7409-8 (pdf) | ISBN 978-0-8071-7408-1
(epub)
Subjects: LCSH: Hoosac Tunnel (Mass.)—History. | Railroads—Massachusetts—
History.
Classification: LCC TF238.H7 B63 2021 (print) | LCC TF238.H7 (ebook) | DDC
385.3/12—dc23
LC record available at https://lccn.loc.gov/2019055134
LC ebook record available at https://lccn.loc.gov/2019055135

For Peggy

CONTENTS

Illustrations follow page 64.

ACKNOWLEDGMENTS

THIS BOOK would not have been possible without the help of several Massachusetts research libraries. Special thanks go to Anna Clutterbuck-Cook at the Massachusetts Historical Society, Beth Carrol-Horrocks and Silvia Meija at the State Library of Massachusetts, and Kim DiLego and Lisa Harding at the North Adams Public Library. I am also grateful to Professor Louis Fereger, of the Boston University History Department, for his advice on Bay State business history and to Charles Cahoon, president of the North Adams Historical Society, for his overall fact-checking and help understanding the casualties at the Hoosac Tunnel. Rand Dotson, editor-in-chief at Louisiana State University Press, took a chance on this book and guided it through publication. James Fallon provided much-needed technical support. My wife Peggy accompanied me to the Hoosac Tunnel several times and encouraged my pursuit of its unique story.

CHRONOLOGY

June 1630   First Puritan ship, *Arabella,* arrives in Massachusetts Bay
                 Colony. John Winthrop declares colony a "City on a Hill."
                 Some fifteen thousand Puritans follow during the decade.

February 1715   European powers conclude Peace of Utrecht, ending naval
                 warfare and allowing Massachusetts shipping to take off.
                 Boston, Salem, and Marblehead flourish.

April 1775   American Revolution breaks out. "Privateering" by Mas-
                 sachusetts ships against British hulls becomes a lucrative
                 wartime industry. Privateer Elias Hasket Derby becomes
                 county's first millionaire.

September 1783   Peace of Paris ends war between England and America.
                 Denied access to British ports, Massachusetts shipping in-
                 dustry crumbles.

August 1784   Boston merchant ship *Empress of China* reaches Canton,
                 initiating rich "ChinaTrade." Elias Hasket Derby's *Grand
                 Turk* soon follows.

December 1807   President Thomas Jefferson imposes an embargo on trade
                 with England. Massachusetts experiences sharp recession.
                 Jefferson's successor, James Madison, declares war on En-
                 gland in June 1812.

December 1814	Treaty of Ghent ends War of 1812. In postwar period, America turns inward. Interior commerce replaces maritime shipping. Massachusetts flounders.
October 1813	Francis Cabot Lowell returns from England with its textile-manufacturing secrets in his head. He founds Boston Manufacturing Company, known as "Boston Associates." He establishes a water-powered textile mill at Waltham, Massachusetts, followed by the larger Lowell mill complex in 1823. The latter employs twenty thousand female workers and is largest factory in western hemisphere.
October 1825	The Erie Canal opens for business. The canal reconfigures American commerce and establishes New York City as the gateway to the nation's heartland. "Erie Fever" sweeps seaboard states, and rival projects are initiated.
Fall 1825	Engineer Loammi Baldwin surveys route of a canal-tunnel through the Hoosac Mountain in the Massachusetts Berkshires. Idea is voted down by the state legislature.
June 1831	Massachusetts legislature approves charter for forty-four-mile Boston & Worcester Railroad, as first phase of a cross-state line connecting to the Hudson River and Erie Canal. Operational by spring 1835, railroad is a major success as a local line.
Fall 1842	The Western Railroad, running the rest of the way from Worcester, Massachusetts, to Albany, New York, is completed. The road is plagued with problems and never coheres with the Boston & Worcester as a viable cross-state railroad to access Erie Canal freight.
August 1843	Alvah Crocker, president of the Fitchburg Railroad, and Elias Hasket Derby III, grandson of the legendary Salem

merchant shipper, travel to Europe. They will ride various railroads and inspect tunnels there. Crocker will become known as the "father" of the Hoosac Tunnel.

May 1848   Massachusetts legislature grants a charter to the Troy & Greenfield Railroad. Along with the Fitchburg, the new railroad will be part of a northern line across the state and challenge the Western Railroad's monopoly. The charter includes a tunnel through the Hoosac Mountain.

Spring 1851   Excavation begins at the Hoosac Tunnel's east portal but is soon stopped due to a lack of funds.

February 1854   In spite of heavy opposition from the Western Railroad, the Massachusetts legislature approves a $2 million loan to the Troy & Greenfield Railroad.

January 1856   Herman Haupt signs on as chief engineer of the Troy & Greenfield Railroad and Hoosac Tunnel. Haupt has served in the same capacity for the Pennsylvania Railroad.

February 1860   The Massachusetts Loan Act of 1860 eases restrictions on the $2 million loan, and work at the tunnel moves ahead. Still, only ten percent of the tunnel's nearly five-mile length has been excavated. The Civil War begins with the attack on Fort Sumter during April 1861.

July 1861   Herman Haupt runs out of money, and work on the tunnel is halted.

February 1863   Governor Andrew's three-man commission issues its report on the Hoosac Tunnel. In accordance with the report, the state takes over the Troy & Greenfield Railroad and the Hoosac Tunnel. The tunnel is estimated to cost $5,719,330. New tunnel plan includes a thousand-foot deep central shaft and a power dam across the Deerfield River. Thomas Doane is hired as chief engineer and work is restarted.

Late 1866   Charles Burleigh successfully deploys his pneumatic drill at the tunnel. A new explosive, tri-nitroglycerin, and fulminate of mercury fuses also come on stream, accelerating excavation of the tunnel.

October 1867   A fire in the tunnel's central shaft building kills thirteen miners and floods the shaft. Tunnel contractors file for bankruptcy. The tunnel is less than 40 percent excavated.

December 1868   Canadians Walter and Francis Shanly agree to finish the tunnel for $4,598,268. A deadline of March 1874 is set. By summer 1869, almost nine hundred men are at work.

October 1869   Torrential rainfall inflicts severe damage at the tunnel's west portal and washes away Troy & Greenfield trackage east of the Hoosac Mountain. The Shanlys recover and carry on.

December 1872   The Shanlys connect the east portal bore to central shaft. This is celebrated as the "first breakthrough." Work shifts to the water-inundated west portal bore.

February 1873   "The Great Debate" takes place in the Massachusetts State House. Issues are whether to privatize the Troy & Greenfield, consolidate the railroads to and from the Hoosac Tunnel, and extend the railroad further west. No decision is reached.

September 1873   Panic of 1873 hits the nation. The ensuing economic depression worst in country's history and lasts for rest of decade.

November 1873	The "final breakthrough" achieved on Thanksgiving Day. The Hoosac Mountain is breached. Walter Shanly requests six-month extension to ready tunnel for train traffic.
February 1875	First railroad train passes through the Hoosac Tunnel.
April 1875	Massachusetts legislature again debates what to do with the Hoosac Tunnel. Decides to go with a "toll-gate plan," calling for state to retain ownership of the Troy & Greenfield Railroad and the Hoosac Tunnel. Toll-gate plan makes both available to any railroad for a fee. Plan does not consolidate various railroads along the Hoosac line and produces insufficient revenue to cover maintenance and service tunnel debt.
February 1887	After declaring the toll-gate plan a failure, Massachusetts sells Troy & Greenfield and Hoosac Tunnel to Fitchburg Railroad. Fitchburg consolidates railroads between Boston and Hudson River and connects to Schenectady, New York. State retains seats on new railroad's board.
Mid-1900	Boston & Maine Railroad leases consolidated Fitchburg Railroad and Hoosac Tunnel. Buys the railroad shortly thereafter. Healthy level of traffic flows through Hoosac Tunnel until 1920s, when a decline of Massachusetts manufacturing undercuts state's commerce. Boston & Maine declares bankruptcy in 1970s but struggles on.
2006	Pan American Railway Company takes over the Boston & Maine Railroad and Hoosac Tunnel. Company manages the tunnel today. Tunnel continues to be used.

BURIED DREAMS

INTRODUCTION

BY AUGUST 1851, Herman Melville knew what kind of book *Moby-Dick* would be. Looking east out the window over his writing desk, he drew inspiration from the looming bulk of Mount Greylock. It was the highest part of the Berkshire Mountains, the unbroken north-south chain that isolated the narrow strip of western Massachusetts where Melville's summer cottage stood from the rest of the state. The shape of Mount Greylock, he wrote a friend, resembled a great Right Whale.[1]

Melville had decided to juxtapose his fictional whale with a maniacal sea captain named Ahab. Melville conceived Ahab after reading British historian Thomas Carlyle, who described nineteenth-century men as "lashed together . . . like rowers on some boundless galley." Modern men, Carlyle argued, were victims of their own technical progress. "We war with rude Nature," he wrote, "and, by our restless engines, come off always victorious."[2]

But such hubris exacted a terrible price, he warned. In Melville's novel, Carlyle's "boundless galley" becomes the whaling ship *Pequod* and Ahab the victim of an obsessive quest to destroy the white whale. Ahab defies his crew to stop him: "Come see if you can swerve me. The path of my fixed purpose is laid with iron rails, where my soul is grooved to run, over unsound gorges, through the hearts of mountains."[3]

As a metaphor for antebellum America, *Moby-Dick* captures the triumphalism of a young, brash, modernizing nation. Allusions to industry, railroads, and the country's burgeoning confidence in its technical progress permeate the novel. While Carlyle worried about his country's transition to modernity, Americans were more positive regarding its impact. Though unsure where it would lead, many Americans associated their nation's rapid technical progress with the drive to expand its boundaries, tame its vast wilderness, and fulfill what they believed was its divine mission to build a better society. Many Americans felt they were escaping the confines of the past and reaching for new, unchartered possibilities. In an 1824 Phi Beta Kappa address at Harvard, Edward Everett, a future governor of Massachusetts, described this exhilarating *Zeitgeist*. "Instead of being shut up, as it were, in the prison of a stationary, or slowly progressing community," he explained, Americans were being "tempted on by a horizon constantly receding before them." Melville conveys this same sense of the sublime in Ahab's fateful quest. "All visible objects are but paste board masks," Ahab declares before his crew. "If man will strike, strike through the mask! How can a prisoner reach outside except by thrusting through the wall?"[4]

It is unclear whether Melville knew, as he wrote these lines, that another crew of men on the opposite side of the Berkshire Mountains were beginning a venture nearly as mad as Ahab's. In a way, they were prisoners too. They were attempting to dig their way out of Massachusetts, by boring a railroad tunnel through the Hoosac Mountain, one of the most imposing escarpments in the Berkshires. By doing so, Massachusetts would open a much-needed gateway to the West, access that region's abundant wealth, and rejuvenate the state's flagging economy. Supporters of the tunnel envisioned western grain flowing through the tunnel to Boston for export to a hungry Europe. In exchange, manufactured goods would travel west to consumers in the heartland eager to purchase them. Americans had never attempted such a feat of tunneling before. At 2,566 feet in height, the Hoosac Mountain would require a nearly five-mile-long tunnel to penetrate it at railroad grade. Some two million tons of rock would need to be excavated from the tunnel. In the early years, this would have to be done by hand drilling. In the end, the tunnel would take almost a quarter century and a terrible loss of life to complete. Although the story of the Hoosac Tunnel has faded from popular memory, it is no less fantastic and emblematic of its time than Melville's novel.[5]

Given the spirit of the time, it was easy to believe in the promise and feasibility of the Hoosac Tunnel. During the mid-nineteenth century, both the nation's dynamic westward expansion and its rapid technical progress helped shape the tunnel's overarching dream and make it seem plausible. The lure of the West was as magical as it was challenging. Its bounty was perceived as virtually limitless but accessible only to those states able to connect to it commercially. The success of New York's Erie Canal demonstrated how richly eastern manufacturers and Atlantic shippers could benefit from access to western markets. The canal's opening in 1825 made New York City the logical destination for western trade and unleashed a fierce competition known as "Erie fever" among rival seaboard states. The so-called market revolution of the 1830s and 1840s intensified the exchange of manufacturing goods for foodstuffs between regions. Once the expansion of the railroads overtook canal building in the 1840s, eastern states with fewer riverine advantages than New York envisioned new and exciting ways to connect with the West. The Hoosac Tunnel was the most ambitious of these schemes.[6]

One historian has observed that not all history can be remembered. What *is* remembered is often done so to bolster the psychic values of a particular nation. American memory has tended to favor those public projects, or "internal improvements" as they were then called, which demonstrated the country's technical prowess, overcame daunting obstacles, and advanced the nation's economic progress. Such projects became "shibboleths of their age" and were enshrined in the public memory. The most famous of these were the Erie Canal, the Transcontinental Railroad, and the Brooklyn Bridge.[7]

So, why hasn't the Hoosac Tunnel found a place in this pantheon of public memory? With construction beginning in 1851 and the first railroad train passing through it in 1875, the tunnel took twice as long to build as the Brooklyn Bridge. Some 195 laborers either died or were severely injured during the tunnel's construction (135 deaths have been verified), making it much more lethal than the bridge. And that's not all. The tunnel was more costly than the bridge. By the time Massachusetts sold off its last tunnel-related assets, the Hoosac Tunnel had drained the state coffers of $28 million. This was an extraordinary sum for its time, equivalent to $875 million today. The tunnel was the state's largest financial obligation until the end of the nineteenth century. Interest payments and upkeep on the tunnel exceeded the state's education budget. Why, then, has the Hoosac Tunnel been forgotten by the public and historians?[8]

The simple answer to this question is that the Hoosac Tunnel never delivered on its grandiose promise of economic rejuvenation for Massachusetts. When the tunnel was completed in 1875, the state did not know what to do with it and eventually sold it off for a fraction of its cost. Although the tunnel enjoyed a degree of commercial success, the shining dream of economic renewal—which motivated the tunnel's supporters to defeat both the forces of nature and their political opposition—proved a chimera. The Hoosac Tunnel, its champions had promised, would allow Massachusetts access to the "cornucopia of the West" and establish reciprocal trade with that region so lucrative it would reverse the state's long and painful economic declension. The tunnel might even allow Boston to challenge New York City as the preeminent *entrepôt* for western goods on the eastern seaboard. Completion of the Hoosac Tunnel became a point of pride for many Massachusetts citizens and support for it a litmus test of their state patriotism. Once it became freighted with symbolic importance, rational arguments against the tunnel's construction became ineffective. Once imbedded in Bay State politics, the Hoosac Tunnel took on a life of its own.

This narrative starts by examining the founding culture and early economy of Massachusetts. It shows how the people of the colony (and later the state) struggled to make up for the poor endowments of their natural habitat with far-flung maritime ventures and bold manufacturing initiatives at home. Again and again, the Bay State reinvented itself economically to compensate for its paucity of natural resources and problematic geography. These economic strategies began to fail, however, when the country's development turned inward after the War of 1812 and the Erie Canal rerouted trade flows away from New England. These broad economic shifts, I will argue, planted the seeds of the Hoosac Tunnel and fueled the determination of many Massachusetts citizens to see it through.[9]

The narrative to follow will also describe how closely the vicissitudes of the Hoosac Tunnel were linked to Massachusetts politics. In light of the number of states that had reformed their legislatures and prohibited publicly funded railroad projects by the 1850s, it is a marvel that work ever began on the Hoosac Tunnel. Work did begin, however, because of both rising anxieties about the Bay State's external competition for western trade, mainly from its nemesis New York State, and intense internal rivalries between parts of Massachusetts for railroad lines to serve their various economic interests. This potent brew of railroad politics allowed work to begin on the Hoosac Tunnel in 1851. But it was stopped shortly thereafter by the rise of the Know Nothing Party, with its antipathy toward state-funded railroad projects. Soon, however, work on the tunnel recommenced with support from the new Republican Party, anxious to build its political capital after assuming governance of the state. After the project went bankrupt in 1862 and was taken over by the state, the Hoosac Tunnel's life seemed assured despite the billingsgate that hounded its glacial progress and soaring cost. The more vehemently its opponents attacked it, the more exaggerated and emotionally charged the defense of the tunnel became. By 1864, there were over six hundred laborers at work on what became known as "The Great Bore." Tunnel politics would dominate the state legislature for another decade.[10]

The narrative includes a walk through the portrait gallery of those individuals who either supported or opposed the tunnel during its construction. These include Alvah Crocker, wealthy paper magnate and undisputed "father" of the tunnel; Herman Haupt, an early chief engineer of the tunnel who later achieved wartime fame as "Lincoln's railroad man"; Edward Hitchcock, the geology professor who "looked inside the mountain" to evaluate the tunnel's feasibility; and Francis Bird, who called the tunnel a "modern Minotaur" and swore to slay it before its insatiable appetite drained the state of tribute. So, too, I will pay homage to common laborers who worked on the Hoosac Tunnel and too often gave up their lives in its darkness.[11]

Special honors are given to the tunnel's chief engineers—in addition to Haupt, men like Thomas Doane and Walter Shanly—who eventually dug their way through nearly five miles of mountain core. These men were required to improvise and invent as they went along. Although the tunnel itself never lived up to public expectations, its construction resulted in impressive technical advances that were by-products of the work on it. Innovations in pneumatic drilling, tri-nitroglycerin, industrial elevators, and other advances in the field of mine engineering filled textbooks and gave rise to whole new industries by the time the tunnel was completed.[12]

Finally, the narrative argues that the public psyche of Massachusetts and the nation shifted during the lengthy construction of the tunnel. The state that began the Hoosac Tunnel in 1851 was not the one that finished it in 1875. After the bloodbath of the Civil War, the grinding depression of 1873, and what was popularly perceived as a moral declension during the final decades of the nineteenth century, the idealism that had given rise to the tunnel and undergirded its promise of economic rejuvenation began to dissipate. Dreamy visions of economic renewal surrendered to skepticism and cynicism. In the end, time caught up to the Hoosac Tunnel. The great railroad boom of the midcentury collapsed from overbuilding, cutthroat competition, and corporate scandal. Transportation patterns had changed since the Erie Canal era, and Massachusetts had reached the point of exhaustion on the tunnel. Denied these vital sources of material and spiritual nourishment, the chimera that had been the Hoosac Tunnel died and was buried with the dreams that had spawned it. When the first train passed through the tunnel in February 1875, the accompanying celebration was dampened by disappointment and a sense of triumphalism miscarried.[13]

1

THE LIMITS OF THE BAY COLONY

NATURE SEEMS TO have had it in for what would become the Massachusetts Bay Colony. More than a billion years ago, the violent collision of continents and microcontinents, some the size of present-day Japan, wrought geological havoc across the region. Over countless millennia, this process worked like a massive accordion, squeezing these land masses together from east and west. This tectonic compression created a wildly jumbled and fractured topography that would prove fateful for inhabitants of the modern era. One result of this process was the western upthrust that formed the Berkshire Mountains, containing traces of the ancient continent of Rodinia in its core. This escarpment would present a formidable impediment to westward travel by the region's eventual inhabitants. It would be through this barrier that the Hoosac Tunnel would have to bore. Farther east, this same process of compression created a rumpled, "corduroy" terrain, with north-south running rivers that would be useless for connecting the region's interior with its eastern coastline. Finally, the ceaseless grinding together of these ancient land masses eroded away any valuable mineral deposits and leached out vital nutrients from the region's surface. Early settlers of this ravaged landscape inherited little more than pulverized granite, sand, and sterile soil.[1]

Glaciation was a more recent phenomenon, covering Massachusetts with over a thousand feet of ice starting a million years ago. Not until sixteen thousand years ago did this incredibly destructive force begin to recede. By this time, successive waves of glaciation had deposited a rocky detritus across the region and, under the weight of their icy mass, embedded this debris in its subsurface. Henceforth, Massachusetts lands would "grow rocks" every spring and require back-breaking labor to clear. This depleted, littered terrain would make subsistence farming difficult and large-scale agriculture impossible for early inhabitants. However, the most prominent feature left behind by glaciation was Cape Cod. This dangerous peninsula south of present-day Boston would account for half the shipwrecks along the continent's east coast. It would pose a lethal hazard to coastal shipping until a canal was cut through it in the twentieth century.[2]

Hemmed in by mountains to the west, harassed by a ship-catcher to the east, hampered by infertile, boulder-strewn fields in between—nature's flawed design did not end there. For all its wealth of coastline, Massachusetts boasts few good harbors and no great inland rivers. Boston Harbor's irregular bottom, numerous islands, and frequent fogs make it inferior to the deep, unobstructed Port of New York. No great rivers such as the Delaware, Hudson, or St. Lawrence permit access to the region's interior. The humble Merrimack River pales next to these majestic waterways.[3]

Nor is Massachusetts blessed with a favorable climate. Temperatures below ten degrees Fahrenheit are not uncommon, though rarely sustained. Severe winter storms, known as "nor'easters," plague the region. The growing season in Massachusetts is short, from May to October, and during the colonial period was even shorter, given cooler temperatures that prevailed. Even when early farmers produced a surplus of potatoes and apples, these crops had little export value because England grew them too. All of this led Oliver Cromwell to declare Massachusetts "poore, cold and useless."[4]

So why were early settlers attracted to this difficult land? Many were misled by colonial promoters, who exaggerated its natural wealth. Some claimed the corn grown by Massachusetts Indians was so sweet and abundant it could be smelled out to sea. Others wrote that deer would come to hunters when called and were easily slaughtered throughout the year. In fact, maize was a difficult crop to grow even with the mounding methods used by Native Americans. Nor was the deer population very large, given the prevalence of wolves in the region. Furthermore, Indians rarely settled in one location for long, as European immigrants tended to do, and depended on their small social units and constant mobility to procure enough food. These indigenous people were much more attuned to the ecosystem and seasonal cycles than were Europeans, suffering from their dangerous misconceptions about the New World.[5]

Geography matters, and the geological events described above influenced the story of Massachusetts. They determined the barrenness of the land, the uselessness of its natural pathways, and its restrictive boundaries. How fortunate that early European emigrants were a determined and innovative breed. They would struggle to make a livelihood in such an austere environment and look for ways to escape its geographical confinement. Without these special qualities, the Bay Colony's early settlers would have neither survived nor prospered in such a harsh and uninviting place.

The very first to arrive famously did neither. Landing at Plymouth in December 1620, the hundred or so Separatists from the Church of England, later calling themselves Pilgrims, bore the brunt of New World realities. Already malnourished and lacking supplies for winter, half died in the first few months. Replenished by new arrivals, the Pilgrims numbered less than three hundred a decade later. Much more successful were the reform-minded Calvinists, better known as Puritans, who settled farther north around Boston and Salem. This so-called Great Migration consisted of some fifteen thousand colonists arriving during the 1630s. Unlike the Pilgrims, the Puritans had remained in the Church of England. They did so with a purpose. Led by visionary John Winthrop, the Puritans meant to reform the established church in England by example from their New World setting. Once the Massachusetts Bay Colony had proven itself, Winthrop and his ministers expected the Church of England to adopt its reformist doctrine and governing principles. Such a bold venture required a large number of migrants to make it a meaningful, strict group discipline to insure its success, and financial support from similarly persuaded elites at home to sustain it. Winthrop probably anticipated some reforming progress in England while he built his colony. Still, he believed the degree of reform achievable amidst the turmoil of the Old World would be limited. Rather, it would be the Puritans' "City upon a Hill," as Winthrop described his colony aboard the flagship *Arabella,* which would influence all.[6]

While naïve about their influence on the mother country, the Puritans would establish a very special colony in the New World. The sheer number recruited for the Great Migration guaranteed their dominance in Massachusetts. Subsuming smaller, kindred groups like the Pilgrims and barring entry of rivals like the Quakers, the Puritans never made religious tolerance part of their agenda. Unlike the Quakers, they believed sinful men and women needed an authoritarian hierarchy of clergymen to keep them in line. In spite of these strictures, the demographic makeup of the Great Migration energized the colony economically and enriched its cultural development. There was nothing spontaneous about the project. Its organizers carefully selected migrants of the "middling sort," with both property holdings and useful skills, to make the crossing to Massachusetts. Migrants sold farms, shops, and other businesses at home to finance their passage and establish themselves anew in the wilderness of the New World. Whole families migrated together, guaranteeing both social stability and a source of labor in their new settlements. Nearly two-thirds of migrants were literate, twice the rate of the land they were leaving. A robust contingent of Puritan ministers, many university educated, accompanied the migrants. Among them were at least 130 alumni of Oxford and Cambridge (mostly the latter). They would provide the spiritual guidance so important to the Puritan project.[7]

The Puritans had many reasons for leaving England. Foremost was their belief that the reformation of the Anglican Church had been sorely incomplete, retaining too much of the hierarchy, doctrine, and ritual of Catholicism. For their part, the Puritans wished to revive the simple church of Christ and his apostles. They believed church authority should be devolved to the community level and, as Puritan scholar Richard Hooker described it, its religious practices simplified to the "bare reading of the word of God." Such an astringent form of worship allowed only preaching and communal meditation on the scriptures.[8]

Central to Puritan theology was the "covenant," a term signifying both the spiritual conversion of an individual to Christ and admission to membership in the Puritan church. While the covenant was largely voluntary, aspiring candidates found attaining it brutally difficult. Going before church elders and the assembled residents of their township, aspirants had to convince them of their profound devotion to Christ and intellectual rigor in interpreting the scriptures. Only those who were successful qualified for leadership positions in the colony, relegating others to second-class citizenry. As early as 1631, the General Court, the colony's highest legislature, declared that only those who had mastered the covenant could serve as representatives in that body. As the colony matured, the covenant proved too over-reaching and exclusionary for the broader community. It would eventually lead to spiritual schisms and political factions that undermined the colony's unity. But during the critical period of its founding, the covenant injected intensity and commitment into the Puritan project.[9]

In addition to purifying England's established church, Winthrop and his ministers intended to correct the economic inequity and social disharmony that had disrupted seventeenth-century England. The unique township pattern the Puritans adopted, for example, encouraged the close-knit communal relationships they sought. As much as possible, civil authority and spiritual leadership were placed in the hands of the townships. Each had its own minister, teacher, town meeting, and selectmen. The careful allocation of property in these townships and strict residency requirements prohibited the growth of a landed aristocracy. Importantly, Puritan settlers owned their land outright as "freeholders," protecting them from burdensome rent increases. Other aspects of the Puritan agenda were directed at balancing the colony's spiritual beliefs with its entrepreneurial energy. The Puritans believed in profitable commerce, seeing it as God's work and their own reward for leading godly lives. They rebuked what they perceived as the greed and rapaciousness of England's merchant and landed classes. Instead, the Puritans' communitarian experiment encouraged a softer edge to commercial dealings and economic enterprise. Fair wages would be paid and interest on loans kept reasonable. Winthrop urged what one historian has called a "covenantal legalism," or informal contractual relations based on neighborly trust, rather than tightly drawn legal agreements. Melding together ambitious shopkeepers, farmers, and artisans—such communitarian attitudes encouraged economic cooperation and shared enterprises that materially enriched the Puritan colony.[10]

The homogeneity of Puritan society and its communitarian attitudes would be critical in meeting the challenges of a land so wanting in natural resources. The earliest settlers to arrive in Massachusetts were quick to acknowledge these shortcomings. "The air is sharp, the rocks many, the trees innumerable," one colonist complained to a relative in England, "the grass little, the winter cold, the summer hot, the gnats in summer biting [and] the wolves at midnight howling." Another colonist described the natural environment as "a terrible wasteland." Yet another lamented "the meanness of the place." Such observations were like alarm bells warning that the Puritan project might fail, if something drastic was not done.[11]

Massachusetts men did not choose to go to sea. They were forced to. Before the end of the 1630s, it became clear that agriculture alone could not sustain the colony. Efforts to improve the poor quality of the soil were disappointing, and, in many places, the soil was already giving out. Much of it was too poor even for pasturing. With the steady influx of new immigrants, the Massachusetts Bay Colony could not feed itself. The sea, however, teemed with abundance. As one farmer-turned-fisherman put it, "We make a shift to live." Faced with desperate economic circumstances, the people of Massachusetts initiated their first great adaptation and established a pattern of innovation into the future. This same drive for economic rejuvenation—to overcome a paucity of natural resources and geological impediments to commerce—would eventually lead Massachusetts men to the Hoosac Mountain.[12]

As they turned to the sea, the Puritans had certain advantages. To begin with, their emphasis on setting up townships quickly had kept them from moving too far inland. They were more or less a coastal colony. What is more, the Great Migration had included many shipwrights and master builders in its ranks. Since the colonists could not afford to purchase new vessels, these artisans set about constructing a fishing fleet suited to local waters. Ironically, the greatest advantage came from the land. Trees covered Massachusetts. They included white oaks, black oaks, cedars, and chestnuts. Some of these species were not just sturdy enough for boat building but also resistant to marine rot. Grandest of all, white pines were perfect for constructing strong, straight masts. Even lowly pitch pines furnished naval stores with pitch, rosin, and turpentine. Then, there were the fish. Most prized were codfish, massive, fleshy, and abundant. Cod could be easily dried, salted, and shipped abroad. It became the lucrative "export crop" the parsimony of the land had denied the colony. "Fish is the only great stapple which the country produceth for forraine parts," a Marblehead resident explained, "and is so benefitiall for making the returns for what wee need."[13]

The efflorescence of maritime Massachusetts is a magnificent and many-faceted story. One innovation followed another. The variety of the colony's water-borne industries and their global reach were remarkable. Still, economic success came with a cost. Over time the wealth generated by these industries and the class divisions they produced undermined the religiosity and social harmony the Puritan founders had fervently sought. By the time the founders passed away (John Winthrop died in 1649), inhabitants of the colony found themselves less attentive to the teachings of their elders and more in the thrall of a booming economy. In short, they had to "learn commerce or perish." The Massachusetts Bay Colony not only survived—it thrived.[14]

First came cod fishing out of the towns like Gloucester and Marblehead. Initially, their fleets worked the nearby Georges Bank and then extended their activity to the much larger Grand Bank off Newfoundland. During the February through September season, larger fishing vessels made three to four "fares" to these cod-rich grounds. During the winter, smaller "coasters" worked less abundant waters closer to shore. The English civil war of the 1640s disrupted competition from the mother country and gave a boost to the Massachusetts fishing industry. The total New England catch, most of it landed in Massachusetts ports, rose from 600,000 pounds in 1641 to six million pounds in 1671. At first, British merchants brought salt to these ports, cured the local catch there, and transported it back to Europe. Soon, however, Boston entrepreneurs took control of this process. Up and down the coast of Massachusetts, the fishing industry spawned related enterprises such as salt works, ropewalks, sail lofts, iron foundries, cooperages, and, most important, shipyards.[15]

Building its own ships became an economic imperative if Massachusetts was going to control the distribution of its fishing catch and maximize profits from it. Furthermore, the English Navigation Act of 1651 stimulated ship building in Massachusetts ports. It restricted trade within the British Empire to vessels of the mother country and its colonies, thereby reducing competition from French and Dutch traders. Since high-quality lumber was the largest component of sailing ships, Massachusetts ship builders exploited their cost advantage over domestic and foreign competition. By the end of the seventeenth century, Massachusetts was supplying vessels to merchant shippers in New York, the Chesapeake, and England.[16]

Most Massachusetts-built ships were absorbed by the colony's own burgeoning shipping industry. One historian has quipped that by going to sea Massachusetts men wished to escape the censorious tone of Puritan society and the scrutiny of its busybody clergy. More likely, shipmasters were motivated by a sense of adventure and the lure of profits. With the Peace of Utrecht in 1715—and the suspension of naval warfare between major European powers—the seas became safer and Massachusetts shipping took off. Boston, Charlestown, and Salem became booming trading ports. In addition to supplying fish to Europe in return for manufactured goods for the American market, Massachusetts shippers opened up a lucrative bilateral trade with the sugar islands of the West Indies. These islands were the most valuable possessions in the British Empire and could pay high prices for fish, lumber, and other provisions. Massachusetts hulls returning from the Caribbean brought back molasses for distillation into rum. By the second decade of the 1700s, Massachusetts boasted sixty-three such distilleries in full production. These ventures whetted the appetite of Bay Colony merchants for new markets. They soon worked their way south to Surinam, Brazil, and Rio de la Plata.[17]

During much of the period from the Peace of Utrecht in 1715 to the American Revolution, peace prevailed. Transatlantic crossings more than tripled, tying Massachusetts more closely to Europe. Migration from Britain actually declined and was replaced by other groups: Scots, Irish, French, and Dutch. (A lessening of civil strife and the onset of industrial development dampened the appeal of British migration.) In Massachusetts, the covenant gave way to the so-called half-covenant. The Puritan church's hold on the colony weakened but did not break. Traditional churchgoers, now referred to as Congregationalists, still fined those who worked on the Sabbath and set the moral tone of the townships. This communal cohesion facilitated investor networks, which was positive for economic development. The resulting concentration of wealth in towns like Boston, Salem, and Marblehead, however, ran contrary to earlier Puritan intentions. By 1770, the richest 10 percent of Bostonians owned 60 percent of the city's wealth. The richest man in Massachusetts was Thomas Boylston, worth an extraordinary $400,000 at the time. He was not alone. Benjamin Pickman of Salem was nearly his equal. In this increasingly stratified society, elite merchants owned the means of production, whether shipping hulls, fishing boats, or whalers. This ruling elite and their attendant ship captains, master builders, bankers, and insurance agents were constantly on the lookout for new ventures to increase their wealth.[18]

However, other seaboard cities soon challenged these merchant elites. By the 1770s, grain exports to southern Europe out of Baltimore were exceeding the value of the Bay Colony's West Indian trade. Furthermore, British manufactured goods were increasingly going to New York and Philadelphia. For its part, Massachusetts had barely enough grain to feed itself and its merchants seemed out of touch with inland retailers. Some conjectured that the Puritans' plain style was out of step with America's rising consumer culture. Others correctly understood that Massachusetts' lack of inland waterways and "Berkshire barrier" hampered its ability to service the developing frontier. The high cost of overland transportation from the colony's ports to the interior made its goods prohibitively expensive. These strains of confinement would intensify over time and create a sense of being hemmed-in regarding access to western markets. Later, these feelings of natural imprisonment would influence arguments in favor of the Hoosac Tunnel.[19]

In the short term, Massachusetts was forced to innovate. Naturally, it leveraged its floating assets. Although the period from the Peace of Utrecht to the American Revolution was generally peaceful, it was not entirely so. And if one includes the Revolution, a different picture emerges. The War of Jenkins' Ear (1739–1748), the Seven Years' War (1756–1763), and the American Revolution (1775–1783) all created opportunities for "privateering." This was a form of state-sponsored piracy whereby fishing and shipping vessels were armed to prey on the maritime commerce of belligerent nations. Whether because of the rough character of its fishermen or the speed of its ships, Massachusetts excelled at privateering. During the Revolution, nearly half of the two thousand or more American privateers sailed out of Massachusetts ports. Its vessels harassed British commerce throughout the Atlantic and supplied American troops with captured munitions. By far the most successful investor in this wartime enterprise was Elias Hasket Derby of Salem. His eighty-five privateering vessels captured 144 British ships of commerce (losing only nineteen of his own fleet). What made Derby successful were his ability to innovate and his force of will. He knew exactly where he fit in the food chain of maritime warfare. He favored smaller, faster privateering vessels that were lightly armed. When much larger British ships-of-the-line spotted him, he piled on extra sails and ran. But when he came upon unarmed British commerce vessels, he pounced. In short, his ships were pint-sized bullies. When he lost a ship, Derby immediately replaced it and then added to his fleet. During the Revolution, Derby made over a million pounds sterling in profits. He became America's first millionaire.[20]

With the Peace of Paris in 1783, the Revolutionary War ended and Massachusetts fell into a catastrophic economic slump. The disadvantages of being outside the British Empire suddenly dawned on Derby and his cohorts. The war had drastically altered old lines of commerce. Gone were the rum-rich Caribbean, trade with Canada, and access to British-controlled ports worldwide. Given its dependence on these destinations, Massachusetts suffered more than other states. It became clear to Derby that the state's merchants needed to wrest themselves from old trading patterns and strike out in new directions. More than anyone else at the time, Derby implemented a new vision for global trade. His ships were the first to carry America's flag to Russia, Ceylon, Burma, Java, and Sumatra. While not the first to fly it in Canton (Samuel Shaw of Boston and his *Empress of China* took that honor in 1784), Derby's *Grand Turk* arrived there six months later. As he had done when privateering, Derby favored small- to medium-size ships, two hundred to three hundred tons burthen, allowing him to safely navigate unmapped waters and never burdening him with more cargo than he could sell. In the early years, he carried a diverse cargo—butter, tobacco, sugar, ginseng, and several thousand Spanish dollars. He never knew what he would be able to sell and what he could purchase in return. Derby was an experimentalist, an innovative genius, who trail-blazed for the many merchants who would follow him. His boldness paid off. By 1800, he had compounded his wealth several times and owned a fourth of Salem's shipping vessels.[21]

While initially casting a wide net, Derby was focused on China by the late 1780s. The British already dominated that market through their East India Company. The popularity of tea drinking in English-speaking society, as well as the appeal of porcelain tea sets to serve it in, made trade with Canton immensely lucrative. Furthermore, both English and American society had developed a seemingly insatiable taste for anything with a Chinese motif—including fine silks, lacquered tea trays, porcelain vases, decorated screens, and even wallpaper. These items imparted status and good taste to their owners and were a part of the rising consumer culture of both countries. The problem for western merchants was that the Chinese wanted few things in return. They preferred Spanish dollars but also accepted a few commodities such as ginseng, pepper, sandalwood, and, oddly enough, dried sea cucumbers. The Chinese considered ginseng and dried sea cucumbers universal curatives with aphrodisiac properties. They used sandalwood to make high-quality furniture and incense. Merchants like Derby could source ginseng from Appalachia but were required to collect the rest of these goods from places like Sumatra, Java, Fiji, and the Hawaiian Islands on the way to Canton. Derby's ship captains were as proficient trading along their route to Canton as they were navigating unfamiliar waters.[22]

Sourcing trading goods for Canton en route was a tedious business, and, given the limited appeal of these goods once landed, they could easily saturate the market. Alternatively, paying for Chinese products with Spanish silver was expensive. British traders solved this problem by supplying the Chinese with opium from India. This gave British merchants a significant competitive advantage. Americans lacked a major trading medium until 1789, when the Columbia, a ship sent out by a group of Boston merchants (among them John Derby, the son of Elias Hasket Derby), picked up a cargo of sea otter pelts at Nootka Sound on Vancouver Island and took them to Canton. This pioneering voyage identified a trading medium which, like opium, the Chinese could not get enough of. The thick, glossy black fur of sea otter peltry appealed to the Chinese mandarin class and was fashioned into hats and other articles of luxury apparel. The voyage of the Columbia kicked off what would become an enormously successful three-cornered trading system for Massachusetts: First, its merchants sailed around Cape Horn to the Pacific Northwest with knives, chisels, and other items to trade for sea otter pelts from the rain-coast natives; then, these merchants carried the sea otter pelts to Canton and exchanged them for tea, silk, porcelain, and other local products; and, finally, they brought these exotic goods back to Salem and Boston. Most of these Chinese products were shipped to other American markets or re-exported to Europe. Soon, merchants in Salem and Newburyport dominated the northwestern fur trade and rivaled the British in what became known as the China Trade. Those predicting the economic withering of Massachusetts had once again underestimated its ability to innovate and reinvent itself in times of crisis.[23]

The era of the China Trade was a high-water mark for the maritime commerce of Massachusetts. Images of harbors bristling with ship masts, tales of enterprising sea captains, and descriptions of boundless wealth pouring into the Commonwealth became imbedded in its collective memory. Later campaigns to reinvent the Massachusetts economy would recall the halcyon days of the China Trade. It had been a dream time for the Bay State and would frequently be romanticized out of all proportion. Elias Hasket Derby's own grandson would harken back to it in numerous speeches in support of the Hoosac Tunnel.[24]

No recounting of Massachusetts' watery endeavors would be complete without some mention of whaling. While Cape Codders taught Nantucket men how to harpoon whales, the latter group fitted out the first whaling ships that could render their catch into its saleable components on board. These so-called tryworks made these ships into floating factories and were in wide use by the 1730s. This maritime innovation allowed Nantucket whalers to cross whole oceans in search of their prey. Over the next hundred years or so the demand for whale oil increased enormously. Before long, Nantucket families like the Macys, Folgers, and Coffins formed an aristocracy that rivaled the elite merchants of Boston and Salem. While American consumers would tire of Chinese knickknacks and the chinoiserie style by the Panic of 1819, they would need whale oil to light their streets and read at night up to the Civil War. (Petroleum was discovered in Titusville, Pennsylvania, in 1859.) Herman Melville would write *Moby-Dick* at his Berkshire summer cottage in 1851, as the whaling industry was peaking and work on the Hoosac Tunnel was beginning.[25]

A second war with England was the last thing Massachusetts wanted. However, the revolution in France, the rise of Napoleon, and a series of Anglo-French conflicts beginning in 1793 shattered the European peace and threatened to embroil America with one or other of the combatants. Still, as long as the Federalists controlled America's foreign policy, the country remained pro-British and pro-peace. While a resentful France harassed its shipping, America fended off its attacks and continued to prosper commercially. This ended when Thomas Jefferson won the presidency in 1800 and the more anglophobic Republicans replaced the Federalists.[26]

Massachusetts had everything to lose in a new war with England. By the first decade of the nineteenth century, it was the largest ship-owning state in America. It controlled over a third of the nation's shipping tonnage and twice that of its nearest rival, New York. Boston and Salem had become world emporiums for pepper, coffee, tea, silk, and Chinese artisanal wares. Furthermore, the Anglo-French wars had added to the state's prosperity. Not only did Massachusetts merchants carry provisions to a war-torn European continent, but they took over many of the belligerents' disrupted trade routes. By President Jefferson's 1807 Embargo, Massachusetts was making over $15 million per year in freight revenue.[27]

While the embargo closed American ports and caused a sharp economic recession, it lasted only fifteen months and was followed by a quick recovery. It was the increased assertiveness of British naval power—and growing anti-British sentiment by the American public in response to it—that pushed the two countries toward war. Although it had surrendered the European continent to Napoleon's armies, Britain was stronger than ever at sea. Its destruction of the French fleet at Trafalgar in 1805 testified to that power and left its navy unrivaled. Britain's depredations of American commerce and its policy of impressment (impressment involved taking American seamen off their ships and forcing them into service on British ships) ultimately brought it into open conflict with America. While the seizure of some four hundred ships hurt American merchants (some were returned after capture), the sharp rise in marine insurance devastated their profits. More serious still, an estimated six thousand American sailors were taken off their vessels and impressed into British naval service. President James Madison, Jefferson's successor, declared war with England in June 1812. However, America was woefully unprepared. Two years into the war, a British ship-of-the-line entered Salem Harbor unopposed. That same summer, the British invaded the Chesapeake Bay and burned the Capitol at Washington. Within months of this calamity, diplomats from both countries were discussing ways to end a war that neither side seemed to have wanted. The Treaty of Ghent in December 1814 officially ended the war and turned back the clock to the *status quo ante bellum*. In terms of territorial gains or settlement of disputes, neither party to the conflict won anything. Nonetheless, Massachusetts was glad to have the war over.[28]

The postwar years were again a disaster for Massachusetts. A pacified Europe quickly reestablished its former trading patterns, and England refused to open its colonial ports to American merchants. Furthermore, English merchants reasserted themselves with a vengeance in the Orient. They made even more effective use of Indian opium procured by their East India Company. Most important, British textile manufacturing was in full gear. It had tapped into an abundant source of raw cotton from India and was producing millions of yards of inexpensive muslin on the steam-powered looms of Manchester and Liverpool. To make matters worse, Latin America was embroiled in its own revolutions, and the Caribbean was infested with pirates. Massachusetts found itself struggling to develop new markets for its carrying trade and, once again, without a trading medium to effectively compete with a rapidly industrializing England. Many Bay State hulls sat rotting in Salem Harbor.[29]

Of course, Massachusetts had faced similar challenges after the first war with England and found ways to surmount them. However, this time would be different. A fundamental shift in America's political and economic orientation altered the ground rules for such a recovery. These changes were heralded by America's lopsided victory over the British in New Orleans at the end of the War of 1812. On a single day in January 1815, the myth of America's divine endowment and exceptional destiny seemed affirmed. At a stroke, the war's conclusion transformed the American psyche. American nationalism ran high, and public policy turned to securing the country from future deprecations from abroad. As part of this postwar reflex, the nation began turning inward politically and economically. The Federalists—and to some extent the New England maritime states—were discredited for their wartime half-heartedness. The frontier, symbolized by Andrew Jackson, the hero of New Orleans, was ascendant. The Federalists went extinct. President Madison and the Fourteenth Congress pursued a nationalist agenda of road building, central banking, and the protection of domestic industry. The lesson coming out of the war was that America needed to be less dependent on foreign commerce and more economically self-sufficient. The Tariff Act of 1816, America's first protective legislation, did not bode well for Massachusetts shippers.[30]

As America's population grew and migrated inland, the South and West increased in political and economic importance. Rising demand for southern cotton and western grain altered both internal and external trade flows. First canal building and then railroad construction would accommodate these trade flows and integrate the nation's economy for the first time. However, the nationalistic fervor of the postwar period would soon give way to regional-interest politics. As a result, the various states in competition with each other would build-out the nation's infrastructure. Furthermore, fluctuations in foreign demand for America's commodity exports and the fragility of the nation's financial system would lead to severe boom-and-bust cycles. Brought on by land speculation, a collapse in cotton prices, and currency inflation, America's first great economic depression arrived in 1819.[31]

Massachusetts' merchant shipping in the postwar years did not disappear. It consolidated to survive. Towns like Newburyport, Marblehead, and Salem saw their carrying business shrink and much of what remained relocated to Boston. Salem's trading fleet declined from 182 hulls in 1807 to 57 in 1815. By concentrating its merchant shipping in Boston, Massachusetts tried to stabilize its trade with the Baltic, Mediterranean, and Far East. However, these relocations did not help much. Boston's commercial tonnage fell by more than 30 percent during the decade that spanned the second war with England. Furthermore, the drastic fall in ocean freight rates, due to unleashed competition from Europe, made shipping less profitable than before. Many Massachusetts merchants gave up their deep-water hulls for coastal trading vessels.[32]

In short, the maritime commerce that had contributed to the halcyon days of the state's prosperity faltered badly and could no longer sustain its economic growth. The Panic of 1819 finished off struggling towns like Salem and plunged Massachusetts and the rest of the country into severe economic depression. After a brilliant start privateering during the Napoleonic Wars, Elias Hasket Derby II, the eldest son of Salem's great merchant-shipping innovator, could not turn a profit on his voyages and sold the family mansion to settle his debts. Frustrated, he took up sheep farming in New Hampshire. During the depression, those with marginal employment and little property suffered worst. A pall of gloom hung over Massachusetts. And yet, a number of the state's merchant-shipping grandees—husbanding their proceeds from better days and ever the innovators—had already begun searching for a new idea.[33]

2

FACTORIES, CANALS, AND RAILROADS

PREVIOUSLY, THE PERIODIC reinventions of the Bay State economy had always involved some new maritime scheme—cod fishing, ship building, privateering, or a far-flung merchant shipping opportunity. Now, Massachusetts would depart radically from this pattern. As America turned inward after the War of 1812 and became less dependent on foreign trade, Massachusetts needed to shift away from the sea for its next reinvention.[1]

So, why did Francis Cabot Lowell, at age thirty-five a prosperous Boston merchant, relocate his family to Scotland in 1810? After settling his wife, Hanna, and their four children in Edinburgh, Lowell traveled around the British Isles for two years. He would return to United States with his family on the eve of the War of 1812. Lowell had grown up in Newburyport and graduated from Harvard in 1793. He was an extraordinary mathematician, earning the praise of Nathaniel Bowditch, America's reigning math genius at the time. After joining his uncle's export-import firm in Boston, Lowell easily mastered the business and rose to become controlling partner. We don't know what Lowell looked like, because he and Hanna never had their portraits painted. They easily could have. By the time he departed for Scotland, Lowell was comfortably wealthy. He told his friends and associates that he was going abroad for health reasons. He kept no journal of his travels and was carefully guarded in his few letters home.[2]

As the second war with England approached, it was clear that the ocean-going commerce of Massachusetts was suffering. Merchant shipping had become more complicated and risky. Despite meticulous planning, one failed voyage could be ruinous. As Lowell left Boston, the venerable merchant firm of Joseph and Henry Lee went under. The business of Patrick Jackson, Lowell's brother-in-law, teetered on the brink. Lowell had tried to diversify by investing in the India Wharf, Boston's second largest harbor-side facility, but its profits were still dependent on merchant shipping. "The hazards of business are much greater now," Lowell observed. What investors needed was a more predictable, dependable form of enterprise which paid steady dividends and required less effort. Lowell set out to find it.[3]

Lowell disclosed the true purpose of his stay in the British Isles to his friend and future business partner, Nathan Appleton, when the two met in Edinburgh. Lowell intended, according to Appleton, to "visit Manchester, for the purpose of obtaining all possible information . . . [for] the improved manufacturing in the United States." In other words, Lowell was an industrial spy posing as a casual tourist. He visited textile factories in Manchester, Lancaster, and other locations observing the latest technology. His superb grasp of mathematics allowed him to put together a detailed mental picture of what he saw on factory floors and learned talking to workers. Since British authorities often searched visitors' luggage if they suspected industrial espionage—they searched Lowell's luggage twice before he left the country—he made no notes or drawings of the machinery he observed. Not since Samuel Slater had pirated the technology for spinning cotton thread from Britain to Pawtucket, Rhode Island, a generation earlier, had anyone attempted such a blatant act of industrial theft. The design for the power loom, invented by Edmund Cartwright several years before and guarded by the British as a national treasure, was capable of producing entire sheets of inexpensive cotton cloth and was the prize Lowell coveted.[4]

Lowell was broadly observant while in the British Isles. He noted the negative effects of industrialism on both the environment and society. He found the manufacturing towns "very dirty" and filled with "beggars and thieves." He came away from England's manufacturing centers shocked by the "debasement of the lower orders of society." In Scotland, Lowell listened to writers like Anne Grant and Walter Scott mourn the passing of traditional folkways and destruction of whole communities in the wake of industrialization. Lowell probably visited New Lanark, Robert Owen's model factory town, and may have talked with its visionary founder. Owen believed factories could both make a profit for their investors and develop the moral character of their workers. Given the similarities between Owen's model factory town and the manufacturing complex Lowell would establish after his return to America, it is almost certain Lowell visited New Lanark.[5]

Lowell wasted no time converting what he had learned abroad into an American reality. He convened the first meeting of the Boston Manufacturing Company, later referred to as the "Boston Associates," in October 1813. Because he had a special vision for the company, he kept the number of stockholders small and firmly under his control. Among its eleven investors, his friend Nathan Appleton and brother-in-law Patrick Jackson were the most important. Lowell also enlisted Paul Moody, a brilliant mechanic with two decades of textile manufacturing experience and numerous patents to his credit, to help him develop the company's first water-powered looms at Waltham, Massachusetts. But Lowell did more than simply replicate what he had seen in British factories. His vision was what one would today call "holistic" and was innovative in several important ways. Lowell wanted industrialism without the negative side-effects he had seen abroad. He wanted to establish a factory system which would leave the social fabric and traditional values of Massachusetts undisturbed. Interestingly, Lowell, Appleton, and several others of the company's investors were descendants of the Puritans' Great Migration during the 1630s. They seemed throwbacks to the Puritan idea that life in the New World could be improved versus the failed experiences of the Old World. Also like their forefathers, the Boston Associates would temper their acquisitive drive.[6]

Lowell's goal was to make sure that industrialism in Massachusetts did not spawn a permanent underclass of factory workers. Instead, he recruited the daughters of rural farm families from across New England as temporary laborers in the Waltham mills. His female workers lived in factory housing, were watched over by responsible matrons, attended religious services, and even published their own newspaper. Although they labored long and hard, Lowell's female workers were encouraged to challenge themselves intellectually in their free time. Recruitment was not a problem, since there was a waiting list to work at Waltham. After several years of service, these women returned to their family farms, took up teaching, or sought marriageable partners, armed with dowries from their factory earnings. Consistent with reigning theories of social improvement and possibly his latent Puritanism, Lowell's factories were "total institutions." Complete control over the lives of his workforce ensured its moral development and spared surrounding society the negative effects of industrialism.[7]

Lowell chose water power versus the preferred use of steam power in Britain to drive his looms at Waltham. This choice was largely dictated by the lack of coal in Massachusetts but also took advantage of the state's fast-running rivers. While useless for commerce, these rivers were one of the few hidden benefits concealed in the area's otherwise poor geography. Water power was far less polluting than burning coal for steam generation, and locating factories on rapid, cascading rivers often distanced them from population centers. In this way too, Lowell's factory system was more benign than its British counterpart.[8]

Lowell's innovative genius exerted itself across the entire spectrum of Waltham's operation. Unlike that of any of his domestic competitors, Lowell's production system was totally integrated such that raw cotton entered one end of the factory and finished cloth left the other end. This allowed the company to capture profit at every step of the manufacturing process and control quality from start to finish. It also allowed Lowell to experiment in order to identify best-selling products. In short, the Waltham mill was America's first example of "mass production." It is estimated by business historians that Waltham was turning out three and a half times the amount of finished product per employee as any other textile factory in the country. As a result of these technological advances, the Boston Manufacturing Company produced cotton fabric competitive in price with coarse British "India cloth" but more durable. America's hard-working farmers and urban laborers quickly came to prefer the sturdier quality of the Waltham product.[9]

Francis Cabot Lowell died in 1817, at age forty-two and only three years after setting the looms of Waltham in motion. One of Lowell's last acts was political. He journeyed to Washington and lobbied for protection under the Tariff of 1816. He convinced Congress to place a protective tariff on imports of coarse cotton fabric, the kind produced at Waltham, leaving rivals who manufactured finer, more expensive cotton goods unprotected. Lowell had the support of southern Congressmen like John C. Calhoun and William Lowndes, who saw the Waltham mills as a promising market for their cotton. Before he died, Lowell had also made plans for a massive scaling-up of the Waltham model at nearby East Chelmsford. The company purchased a three-hundred-acre tract of land on the Merrimack River, where Paul Moody designed a thirty-foot-high waterwheel to capture the powerful descent of the river. Nathan Appleton and Patrick Jackson supervised the construction of six brick buildings for a new contingent of female workers. The new complex, renamed "Lowell" in honor of the great innovator, was landscaped with gardens and neatly laid-out pathways. It included a church, a library, and other such facilities for its employees. The factory opened in 1823 and would ultimately house twenty thousand workers. By 1826, Lowell was the largest industrial complex in the western hemisphere and returned a steady 20 percent dividend to its investors. The company expanded its product line to include specific articles of clothing and various grades of fabric. Distribution penetrated deeper into the South and West. Export sales took off as foreign markets acknowledged the superior quality of American textiles over British goods. China became a major market for the company.[10]

But Lowell had done more. He had redirected the capital of Massachusetts merchant shippers into textile manufacturing. Investors like Abbot and Amos Lawrence, William Sturgis, and Harrison Gray Otis gave up their old businesses and became rich following Lowell's lead. The Boston Associates gradually expanded their investor base, and competitive firms copied their innovations. So, too, those politicians who had eschewed the notion of domestic manufacturing became ardent boosters. Massachusetts senator Daniel Webster was one of its most fervent promoters and helped pass the important Tariff of 1824. In fact, protectionism would become one of the main tenets of the nascent Whig Party and a driving cause célèbre of party leaders like Daniel Webster and Henry Clay. By introducing a more benign rendition of industrialism to America, Lowell had made it respectable to the country's exceptionalist sensibilities. By 1833, Massachusetts congressman Edward Everett would claim that American industry bore no resemblance whatsoever to the "infernal mills of Europe." Likewise, Everett's friend and archprotectionist John Pendleton Kennedy argued that "we have no class in America corresponding with the operatives, the human machines, of Europe. . . . Labor is not with us the mere instrument of capital." For his home state of Massachusetts, Lowell had breathed new life into a floundering economy. He had found yet another way to reinvigorate a place so lacking in natural endowments and seemingly without hope of economic resuscitation. During the 1830s, Massachusetts would become the preeminent state for textile manufacturing. Once again, its remarkable ability to innovate had overcome its paucity of God-given resources. In an 1835 speech, Lowell's son would capture his father's immense contribution and its significance for Massachusetts. "The prosperity of my native land," John Lowell declared, "which is sterile and unproductive, must depend . . . first on the moral qualities and secondly on the intelligence of its inhabitants." The rhetoric of economic rejuvenation was by now imbedded in the civic culture of Massachusetts. Its tropes and metaphors would reappear in the debates over the Hoosac Tunnel by midcentury.[11]

Americans celebrating the country's fiftieth birthday on July 4, 1826, could not have imagined the economic boom they were about to enter upon. After all, the year-long Jubilee festivities—with the nationwide tour of Revolutionary War hero Marquis de Lafayette, the seemingly divine summoning of both Thomas Jefferson and John Adams to their maker on the very day of the country's anniversary, and the countless speeches, sermons, and parades honoring the wisdom and courage of the founding generation—all looked backward to America's beginnings. And yet, in spite of this celebratory nostalgia, the country was modernizing rapidly with the completion of the Erie Canal and the Lowell textile mills. Both were seen by Jubilee celebrants as great accomplishments of their own generation and, in fact, had laid the foundation for the economic boom of the 1830s. The industrialization of Massachusetts and the opening of the West through New York State would be important factors in the nation's economic expansion.[12]

The Erie Canal was an old idea. Albert Gallatin, President Washington's secretary of state, had recommended it to his boss. After the federal government refused to finance such a costly venture, New York State decided to go it alone. Pushed hard by its indefatigable advocate and soon-to-be governor DeWitt Clinton, the New York legislature approved the canal project in 1817. In his memorial recommending the canal, Clinton predicted that the Erie Canal would transform New York City into "the great depot and warehouse of the western world" and "create a new era in history." There was no debate about the canal's path. Nature had defined it. From Albany on the Hudson River, through the Mohawk Valley, to Buffalo on Lake Erie, the highest point above the Hudson was only 650 feet. It was a geological gift to New York State. Furthermore, there was enough water from each end of the 364-mile canal to adequately flood it. Though there were few professional engineers in America at the time, the canal was a masterpiece of engineering. There were eighty-three locks to raise and lower boats and eighteen aqueducts to carry them over rivers and other impediments. Two of these aqueducts, needed to cross the Mohawk River near Schenectady and Cohoes Falls, were 748 and 1,188 feet long, respectively. The cost of the Erie Canal was somewhere in the neighborhood of $4.9 million, or $13,400 per mile. This was less than its builders had expected and less than many later canals. Even before the canal officially opened in October 1825, its completed sections were crowded with barge traffic. In its first year, the canal carried more than a quarter of a million tons of merchandise and over forty thousand passengers. Freight rates from Buffalo to New York City fell from $100 to $5 per ton. The Erie Canal reconfigured American commerce by joining two great regions of the country, the East and the West. It established New York City as the gateway to America's developing heartland and the nation's most important port. The Erie Canal's success stimulated a nationwide frenzy of canal building.[13]

In spite of its commercial success, the Erie Canal came with a human cost. By the time it was fully operational, some thirty thousand men, women, and children worked on the canal. Boys as young as six, but usually eleven or twelve, drove mules and horses along the canal's towpaths and pulled boats behind them on towlines. Locals called these boys "hoggees." Many were orphans, and few received any schooling. One observer thought they were the most profane youths anywhere. Young girls served as maids on passenger boats but often drifted into prostitution. Many were released from their jobs when they became pregnant. Adult males who steered the canal boats and did heavier jobs were called "canallers." With an estimated 1,500 grog shops along the canal, four for every mile of its length, canallers often drank while waiting to pass through locks or during other idle moments. Boat teams often fought with each other for position at the locks, and accidents were frequent. Most canallers did not know how to swim and some drowned in the seven-foot-deep canal when drunkenness led to carelessness. Herman Melville, who spent his honeymoon on the canal in 1847 and may have worked there during the summer of 1840, describes a canaller in his novel *Moby-Dick*. He has a "brigandish guise," Melville writes, "his slouched and gaily-ribboned hat between his grand features. [He is] a terror to the innocence of the villages through which he floats." Middle-class residents of canal towns considered canal workers the lowest class of society. A local newspaper claimed that a quarter of the inmates from Auburn

State Prison had gone to work on the canal. In fact, many canal workers were Irish, English, and Scottish immigrants, the same groups that would later find their way to the Hoosac Tunnel.[14]

For all of the money spent and effort expended, America's so-called canal era lasted only a dozen years. It began in earnest with the inauguration of the Erie Canal in 1825 but was badly disrupted by the Panic of 1837. During the 1840s and 1850s, the coming of the railroads dealt a death blow to most American canals. Even the Erie Canal experienced the gradual erosion of its traffic to the New York Central Railroad. In the end, most canal projects never came close to achieving the financial success of what DeWitt Clinton had done with the Erie. Canals were expensive, costing $20,000 to $30,000 per mile versus the best turnpikes at $5,000 to $10,000. Some canals cost much more and were financial disasters. Engineering at the start of the canal era was in its infancy and many canals were poorly designed and costly to maintain. Furthermore, a lack of water in dry months, flooding during rainy periods, and freezing conditions in the winter could terminate service. Some canals should never have been built in the first place. Local enthusiasm and political logrolling often outweighed common sense in canal building.[15]

Nonetheless, Americans dug 3,326 miles of canals during the 1820s and 1830s. Begun in 1826 and finished in 1834, Pennsylvania's "Main Line" Canal was probably the country's most ambitious canal project. From Philadelphia it followed the Susquehanna and Juniata Rivers to the Allegheny Mountains, where a series of inclined planes and cable cars lifted canal boats up over peaks 2,200 feet above sea level and down the other side. From there, the canal utilized the Conemaugh and Allegheny Rivers, passing through 174 locks, to reach Pittsburgh. The Main Line Canal cost over $12 million but did a decent business. However, its elaborate inclined plane–cable car system and excessive lockage made the canal too slow and cumbersome to compete effectively with the Erie.[16]

The Chesapeake and Ohio Canal was another ambitious project but with an older history. As president of the Potomac Company, George Washington had been the idea's primary booster but had watched work stall at the Potomac Rapids above the nation's capital and eventually succumb to cost overruns. Taken up again in 1828, the Chesapeake & Ohio Canal did not reach Cumberland, Maryland, until 1850 and never made it to the Ohio River. By that time, the

Baltimore & Ohio Railroad ran along the same route and made the canal redundant. The canal cost upwards of $11 million and drove the State of Maryland to the verge of bankruptcy. Although useful delivering coal from the area around Cumberland to the Port of Baltimore, the Chesapeake & Ohio Canal never paid for itself.[17]

Ironically, Massachusetts was a pioneer in canal building but never committed to it as an infrastructural strategy the way other states did. Begun in 1794, the twenty-seven-mile Middlesex Canal eventually connected Boston Harbor with the Merrimack River at East Chelmsford (later renamed Lowell). Bay State elites like John Hancock, John Adams, and Christopher Gore purchased shares in the canal company. Loammi Baldwin, later known as the "father" of American civil engineering, surveyed the canal and designed its structure. Baldwin's innovations included a new form of water-tight cement for sealing the stone locks and a precursor of the dump truck for more efficient earth moving. Engineers on the proposed Erie Canal inspected Baldwin's work before breaking ground on their own project. Opened in 1804, the Middlesex Canal connected the interior of Middlesex County more directly with Boston Harbor than going down the dog-legged Merrimack River to Newburyport, at the river's mouth, and then south along the coast to Boston. Once the canal opened, Newburyport was bypassed and suffered economically. After the establishment of the Lowell textile mills, the canal served as a conduit for raw cotton arriving at the mills and finished cotton fabric leaving them. Still, the Middlesex Canal was expensive for its day, costing $592,000 to build, and it returned a paltry 2 percent in dividends. It soured the appeal of similar projects in Massachusetts for years to come.[18]

Other Massachusetts canals were even less successful. The Blackstone Canal was intended to end the economic isolation of Worcester, Massachusetts, by connecting it to Providence, Rhode Island. The canal used the Blackstone River, a number of nearby rivers and ponds, and a complex series of locks to span its forty-five-mile length. The canal was poorly designed and badly built. It suffered from low water on the Blackstone River during part of the year and poor maintenance during most of its troubled history. The Boston & Worcester Railroad, operational by 1835, seriously diminished the canal's traffic. The eighty-six-mile New Haven and Northampton Canal was the longest canal in New England and suffered from similar problems. It was plagued with low water and a lack of operating funds. It cost over a million dollars to build but was not ready for business until the mid-1830s. By that time, many of its wooden locks had rotted out. The canal closed within a decade due to a lack of traffic and was filled in as a railroad bed.[19]

The most ambitious Massachusetts canal project never made it off the drawing board. Initially proposed in 1819, a "tunnel-canal" through the nearly five-mile-wide, 2,566-foot-high Hoosac Mountain would have connected the Deerfield River on the mountain's eastern side with the Hoosic River on its western side. This seemingly fantastical scheme was a direct response to the challenge posed by the Erie Canal, then under construction. It was intended to open a route through the state's "Berkshire barrier" to the Hudson River and the Great West beyond. As the most knowledgeable canal builder of his day, Loammi Baldwin was commissioned to survey the proposed route in 1825 and judged the tunnel-canal feasible. "There is no hesitation in deciding in favor of a tunnel," he declared. Still, the tunnel-canal proposal was hotly debated in the Massachusetts legislature and repeatedly voted down because of its uncertain cost and feasibility. It did not help that the canal route would have bypassed the state's three largest cities west of Boston: Worcester, Springfield, and Pittsfield. These legislative delays bought the state enough time that emerging railroad technology seemed to offer a more cost-effective way to breach the state's western barrier. While little was known about this new technology, some believed a railroad line might be able to cross over the Berkshires at a lower elevation south of the Hoosac Mountain. Such an alternative would avoid tunneling through the Berkshires altogether. Although the state would eventually opt for this seemingly more sensible plan, Massachusetts was by no means done with the Hoosac Mountain. Two decades later, the state would return there with a daring new scheme for a railroad tunnel through the Berkshire barrier. Before that, however, the saga of the Western Railroad needed to play out.[20]

For Americans celebrating the opening of the Erie Canal and the nation's Jubilee in the mid-1820s, railroads were still an abstraction. However, rumors of their development were exciting and promising. Some thought they might offer a way to revive the commerce of states without access to great rivers or other inland waterways. In his 1825 inaugural address, Massachusetts governor Levi Lincoln, Jr., focused on the subject of internal improvements to include both canals and "another mode, railways, that have been approved of in England." Lincoln was a native of Worcester and sensitive to the transportation needs of the state's interior communities. Like most Americans, his information about railroads was sketchy. After all, he had never actually experienced one. He had never heard a locomotive's shrill whistle, never felt its reverberations underneath his feet, nor been enveloped in its hissing blast of steam. Somewhat presciently, he wondered how a railroad's tracks "would be affected by our severe frosts." In his address, Lincoln went on to point out that the state was losing commerce to its neighbors. "The most serious diversions of trade are taking place to other markets," he warned. Bold ideas were needed to restore the Bay State's competitiveness, and, he suggested, perhaps this new form of transportation might be one of them. Elias Hasket Derby III, grandson of the great Salem merchant and a future railroad lawyer, agreed with Lincoln. "Gloom and despondency seemed to settle upon Massachusetts," Derby observed, "[and] grass began to invade the wharves and pavements of her commercial centers." Never at a loss to venerate its past glories and warn of its imminent decline, Derby observed that Massachusetts "seemed to stand at the ancestral tomb, sorrowing that she could not partake of the progress of the age." Derby would be one of the lead actors in the turbulent drama of the Western Railroad. He would also become a key booster of the Hoosac Tunnel.[21]

At the time of Governor Lincoln's inaugural address, British railroads were still in their infancy. However, steam-powered technology, perfected by Scottish inventor James Watt, was well-developed across the country. Initially used to pump water out of mine shafts, it quickly gave rise to steamboats on both sides of the Atlantic. By the 1820s, thanks to American inventors John Fitch and Robert Fulton, steamboats were plying the Mississippi, Great Lakes, and Hudson River. Some of the success of the Erie Canal was due to the steamboat link between the Port of New York and Albany, fully operational by the canal's opening. The more challenging concept of a steam locomotive riding iron rails across uneven terrain took longer. When it opened in 1825, Britain's Stockton & Darlington Railroad was no more than a crude prototype of a modern railroad but probably inspired Massachusetts governor Lincoln's comments that year. Four years later, a gruff, self-taught British genius named George Stephenson took a great leap forward with his locomotive *Rocket*. It was fast for its era and dependable over short distances. It proved that railroads could supply efficient overland transportation. Beginning operations in 1830, Stephenson's

Liverpool & Manchester line became the world's first functioning railroad and a testing ground for future locomotive development.[22]

The transfer of railroad technology from England to America was more challenging than it had been for textile manufacturing or steamboats. Railroad construction was more complicated and expensive. Every railroad was different. Land had to be acquired before tracks could be laid. Various roadbeds and rail types needed to be tried out, depending on climate and topography. Sufficient locomotive power had to be developed for different gradients and freight loads. Initially, the British had the advantage with better engineers, an abundant supply of iron, and freer access to capital. They tended to build more solidly than Americans with the view that railroads would last for the ages. Although American inventors traveled to England to learn what they could from George Stephenson's locomotive trials, their early ventures with the Charleston & Hamburg Railroad in South Carolina, Camden & Amboy in New Jersey, and Baltimore & Ohio in Maryland were fraught with problems and suffered occasional disasters. In the fall of 1833, a train wreck on the Camden & Amboy injured two dozen passengers, among them future railroad magnate Cornelius Vanderbilt and former president John Quincy Adams. Nonetheless, American railroad builders persevered and had several advantages to help them. One was the right of "eminent domain" in U.S. law, allowing railroads to purchase property at lower cost than in Britain. Another was the 1832 elimination of the tariff on imported iron if used for railroad tracks. As with earlier ship building, America's abundant forests also proved a blessing. Wood was less expensive and more available than coal to fuel locomotive engines. Finally, the vast size and rapid settlement of the American continent elicited support for railroad construction from many state legislatures. By 1835, there were almost a thousand miles of railroad track in the United States.[23]

Given their unique requirements, American railroads became distinctive. For example, Robert L. Stevens, engineer for the Camden & Amboy, designed the inverted T-rail to replace the strap-iron rail. The new rail was easier to install and held up better under America's larger locomotives and heavier loads. American railroad builders soon learned that wooden cross-ties on gravel roadbeds were less vulnerable to frost heaves than British roadbeds of granite blocks. Similarly, John B. Jervis solved the problem of tight curvatures with his "bogie" wheels on the front of American locomotives. Other innovators made the eight-wheel locomotive with its capacious cab and "cowcatcher" standard in the New World. Large kerosene-burning headlights and piercing steam whistles soon rounded out the form and function of American locomotives. By the mid-1830s, corridor-type passenger coaches were used on many American railroads. Many saw them as a democratic rejection of compartmentalized British coaches.[24]

The Baltimore & Ohio Railroad was America's first major railroad. It was very early. Incorporated in Maryland during 1828, its organizers had no idea what kind of locomotive or rolling stock would run on its tracks. Fortunately, Peter Cooper, a glue manufacturer from New York and inventive genius with multiple patents to his credit, began testing locomotive prototypes on a stretch of track between Ellicott's Mills and Baltimore. Cooper's compulsive tinkering with a mechanical musical cradle, early dishwasher, and lawnmower would not land him in the history books. His work on the locomotive would. By 1830, Cooper's tough little engine *Tom Thumb* was performing as well as Stephenson's *Rocket* had in England the year before. Largely because of Cooper and his fellow inventors, Baltimore became the cradle of American locomotive innovation. So, too, the Baltimore & Ohio became a laboratory for railroad engineering. Not only was the railroad's 380-mile length unprecedented—reaching Cumberland, Maryland, in 1842, and Wheeling, Virginia (now West Virginia) a decade later—but the rugged terrain the line traversed threw every challenge in front of its engineers. As it worked its way westward, the Baltimore & Ohio achieved one construction superlative after another. The bridge over the Patapsco River near Baltimore was the longest in the nation. The Kingwood Tunnel near Fairmont, Virginia (now West Virginia), was also the longest in the country. The chief engineer overseeing these projects was Benjamin H. Latrobe, Jr. He was the son of Benjamin Henry Latrobe, America's most renowned architect and rebuilder of the United States Capitol after the War of 1812. The young Latrobe would later serve as a consultant to the Hoosac Tunnel.[25]

Early American railroads were clearly a work-in-progress. This was certainly true of the first railroad across Massachusetts, the Western Railroad. Early enthusiasm for the Western exceeded any conception of how to build it. Nathan Hale, named for his uncle of Revolutionary War fame and editor of the *Boston Advertiser*, tirelessly promoted the idea of a railroad connecting the state capital with the Hudson River and gained the ear of Governor Levi Lincoln. Many suspected that Lincoln's speeches on the Western Railroad were written by Hale. Theodore Sedgewick, who represented economically isolated Berkshire County in the state legislature, and Harrison Gray Otis, mayor of Boston and a member of the state's industrial elite, whipped up support for the Western Railroad from opposite ends of the state. Abbot Lawrence, a wealthy investor in Boston Associates, was also an important backer. James F. Baldwin became the railroad's chief engineer. He was the fourth son of Loammi Baldwin, who had surveyed the proposed tunnel-canal through the Hoosac Mountain a decade earlier. In contrast to his father, the younger Baldwin proposed a more southerly route across a lower section of the Berkshires. The Western Railroad would reach the Hudson River by way of Worcester, Springfield, and Pittsfield, Massachusetts, and terminate across the river from Albany, New York. Such a route would require little of the tunneling his father had proposed and, the younger Baldwin estimated, cost no more than $3.2 million.[26]

Still, nothing provoked more intense political combat between various parts of Massachusetts than a project of such magnitude and the threat of higher taxes to pay for it. The battle for the new railroad line became the central issue in the 1829 state election. The *Boston Courier* took the side of aggrieved coastal communities such as Newburyport, Salem, and the towns of Cape Cod and called the project "as useless as a railroad to the moon." The paper accused the railroad's promoters of sloppy accounting and naïveté regarding the Western Railroad's true cost and feasibility. The cost per mile of the proposed line was less than two-thirds of its British models, the *Courier* pointed out, and would traverse much more challenging terrain. The paper reminded readers that neither the Middlesex Canal nor any other internal improvement in the state had ever paid back investors as promised.[27]

Faced with such strong opposition, Hale, Otis, and other railroad supporters staged a tactical retreat. After several defeats in the legislature, they proposed a shorter railroad as a first phase of their more ambitious plan to span the state. The single track, 44.5-mile Boston & Worcester Railroad would cost $600,000 and be privately funded. It would serve as an "experiment" to convince the public that a cross-state railroad was viable. It was anticipated that passenger traffic alone on this busy corridor would make the shorter railroad immediately profitable. In June 1831, the Massachusetts legislature approved the Boston & Worcester charter.[28]

Near term, the Boston & Worcester was a great success. To begin with, the railroad's charter was a legal masterpiece. While paying lip service to the public interest, it established monopolistic control over an important area of the state and gave the railroad complete freedom to set passenger fares and freight rates. To raise capital, the railroad cast a wide net, offering ten thousand shares at $100 par value per share. This allowed large and small investors to participate in the stock offering. After an initially tepid reception, important Boston financial houses such as John E. Thayer & Brother and the Massachusetts Hospital Life Insurance Company jumped aboard. Small investors soon followed. So, too, the Boston & Worcester overcame significant construction hurdles during what was an early period of railroad building. Like the Baltimore & Ohio, the line had little experience to draw on. The company secured the right of way for its tracks but at a higher cost than anticipated. After winter frost heaves destroyed its first set of tracks on granite blocks, the railroad installed a foot-deep gravel roadbed with wooden cross-ties that held up better. Initially, locomotive selection was hit or miss. Several British engines failed until new ones from Mid-Atlantic and New England machine shops became available. Overall cost overruns were in the neighborhood of 50 percent. Nonetheless, the Boston & Worcester Railroad opened for business in the summer of 1835 and by its third year—once operational problems like snow clearance, railroad-crossing safety, and scheduling issues were worked out—flourished. On its first Fourth of July promotion, the line carried 1,500 passengers on four different trains. For those living in east-central Massachusetts, the railroad age had arrived.[29]

Viewed longer term, however, the Boston & Worcester took on characteristics that undercut the larger vision of an efficient through-line across the state. One mistake was choosing Boston's South Cove as the line's eastern terminus. This location was the least expensive for the financially constrained company but could not be reached by deep-water ships. Terminating at the South Cove necessitated expensive cartage to access better-suited wharves to the north. Another problem was the Boston & Worcester's focus on passenger traffic. This emphasis contradicted the original concept of a freight line from the Hudson River to Boston Harbor. By the time work began on the so-called Western Railroad—intended to run from Worcester to the Hudson River—the Boston & Worcester had become a highly profitable, local passenger railroad with stockholders more interested in receiving their dividends than being part of a freight-hauling, through-line across the state. This was not just a Massachusetts problem. The story of shorter, local railroads unwilling to consolidate in order to build long-haul, through traffic played havoc with the development of railroads around the country during the 1840s and 1850s.[30]

Although the state granted it in 1833, the Western Railroad's charter lay dormant for four years. Time was needed, of course, to complete the Boston & Worcester. Also, the period from 1830 to the financial panic of 1837 was one of unprecedented prosperity for Massachusetts and the nation. The value of the nation's cotton production tripled and created a multiplier effect throughout the economy. Land sales quadrupled, brought on by demand for more cotton acreage in the South and farmland to grow foodstuffs in the West. Massachusetts textile production soared. The coarse cotton fabric from the Lowell mills was perfect for clothing the South's burgeoning slave population. For the moment, the state's wealthiest investors put aside their concerns about interstate commerce and transportation infrastructure. They were too busy making money.[31]

Another reason for procrastination on the Western Railroad was the well-founded suspicion that its construction would be a daunting task. With 156 miles between Worcester and the Hudson River, the new line would be three and a half times longer than the Boston & Worcester. A new bridge over the Connecticut River, strong enough to support a fully loaded freight train, would be a project all its own. But even before it reached the river, the railroad faced gradients west of Worcester exceeding sixty feet per mile. (The maximum gradient between Boston and Worcester was thirty feet per mile.) The region's tectonic compression a billion years earlier was to blame for this. The Charlton Ridge rose 430 feet just a few miles out of Worcester and required a forty-foot cut through hard rock to get over it. After that, the terrain dropped almost nine hundred feet to the town of Springfield on the Connecticut River, an elevation nearly the same as Boston. Still ahead were the Berkshire Mountains. Even though the Western's route took advantage of a lower elevation than the earlier Hoosac tunnel-canal proposal, there was nothing remotely like New York's Mohawk Valley, the "magic" pass that had made the Erie Canal feasible. Instead, the Western's route rose 1,400 feet above Springfield, requiring its engineers to deal with gradients of eighty feet per mile. To take advantage of what river valleys existed, the Western was forced to construct bridges with spans of sixty feet and as high as seventy feet above the riverbeds. Going up the

Pontoosac River on the eastern slope of the Berkshires, the railroad bridged the river twenty-one times in the course of thirteen miles. In addition, deep cuts, embankments, and a tunnel of 548 feet were needed to provide passage for the railroad. Along the summit at Washington, Massachusetts, engineers made a half-mile cut fifty-five feet below ground through solid rock. Even with these modifications to nature's great barrier, any train traversing the route would require a technological leap in locomotive power. All of this work and equipment was hugely expensive and only gradually understood by the railroad's supporters, the public, and politicians in the Massachusetts State House.[32]

Ironically, while the nation's economy boomed during the 1830s, America's politics descended into chaos. Assuming the presidency in 1829, Andrew Jackson proved to be a force of change, most of it directed at reducing the role of the federal government and putting an end to its suspected collusion with the nation's business interests. First came his veto of the Maysville Road Bill in 1830. This blow to federal support for internal improvements—following President James Madison's similar veto a dozen years before—ensured that the country's great infrastructural projects would be relegated to the various states in the foreseeable future. Then came Jackson's "bank war," beginning with his veto of the rechartering of the Second Bank of the United States in 1832 and culminating, the following year, with the transfer of its assets to his "pet banks" in the states. These actions destroyed the nation's most powerful financial institution and inflated the currency as the pet banks began printing their own money. Soon the nascent Whig Party was vilifying Jackson as "King Andrew the First" for his seemingly autocratic and capricious behavior. The Whigs became a powerful counterforce to the Democratic Party's radicalism and their leader's dismantling zeal. In Massachusetts, multiple-term Governor Levi Lincoln held together the Old Federalists and nationalist wing of the Democratic-Republicans and helped guide these conservative elements into the Whig Party. The Whigs differed little, Lincoln declared, from the "high objects of general improvement and public good" espoused by those earlier parties. The Boston Associates became staunch Whigs. Men like Nathan Appleton, Abbot Lawrence, and Patrick Jackson supported a high protective tariff, restoration of the national bank, and public funding for internal improvements, all core tenets of the Whig Party. Daniel Webster would represent these men in the U.S. Senate for a quarter century and became their chief spokesman on the national stage. Edward Everett, former minister, Harvard professor, and Brahmin politician, became their governor in 1836. The communitarian ethic and social cohesion that had typified Massachusetts's power elite during its earlier history now embraced the Whig Party. If there was a single reason the Western Railroad received financial backing from the State of Massachusetts, it had to do with the state's powerful Whig regime and the support of Webster, Everett, and the wealthy constituency they served.[33]

Without that kind of support, the Western Railroad would never have been built. It was anticipated as an expensive project to begin with, and Governor Everett signed a bill making the state a part owner in the railroad in 1836. After their long struggle for state assistance, advocates of the Western Railroad were ecstatic. Soon after construction began, however, the Panic of 1837 hit Massachusetts and the nation like a financial earthquake. It was caused by a reckless printing of banknotes by Jackson's pet banks, overspeculation in land, a collapse in cotton prices, and a tightening of credit by British banks. It kicked off the worst depression in the country's history up to that time and lasted into the 1840s. Western Railroad stockholders lost an estimated $20 million of their total assets in the depression and could not fulfill their stock pledges to the railroad. At the same time, the railroad's engineers continued to raise the estimated cost of construction. With costs escalating and the state forced to increase its aid, no one bothered to calculate the railroad's rate of return. By its completion in 1842, the Western Railroad had spent an extraordinary $8 million (the Boston & Worcester had cost $1.7 million). Of this amount, $5 million was held in bonds by the State of Massachusetts and the City of Albany and required interest payments of $310,000 per year. This amount exceeded the Western's entire operating budget. Massachusetts also held $1 million in the railroad's stock. While stock dividends would be defrayed until the mid-1840s, the fixed interest on the railroad's bonds had to be paid every year. This obligation strangled the Western Railroad financially and meant it could never charge rates low enough to attract the Erie Canal freight for which it had been built.[34]

In short, the Western Railroad never stood a fighting chance. Besides being saddled with debt, it was too large, too early, and stuck in a bad corporate marriage to the Boston & Worcester Railroad. In the early 1840s, the Western was one of two very large railroads in America. It was exceeded in length only by the Baltimore & Ohio but traversed the more rugged terrain of the Berkshire

Mountains. Like the Baltimore & Ohio, the Western had yet to develop the organizational structure and professional management needed to run a large railroad. No American manufacturing company—neither the Lowell mills nor any other factory operation at the time—faced the same degree of business complexity as these large, early railroads. As a result, the Western Railroad made every mistake possible during its "learning" years. It suffered a series of horrific train wrecks before timetables were enforced and safety procedures implemented. So, too, the company purchased a fleet of untested, shoddy engines that repeatedly broke down and were ultimately junked as a costly failure. Similarly, the railroad made bad decisions at both its western and eastern terminuses. In Boston, it delayed too long replacing the Boston & Worcester's poor South Cove terminal with wharfage offering deep-water access. In Albany, the company tried several strategies for transferring freight arriving on Erie Canal boats to its railroad yard across the Hudson River. Western freight customers complained bitterly about long delays and logistical confusion until a bridge was built across the river in 1866. Overshadowing all of these problems was the strained relationship between the Western Railroad and the Boston & Worcester Railroad. Where cooperation was essential to carry freight between Albany and Boston, the two companies often behaved like hostile competitors. The Western depended on hauling grains and other commodities and, given the low population density of the area it served, had much less passenger traffic than the Boston & Worcester. For its part, the latter railroad had little interest in the Western's freight business. Given its lucrative monopoly over rail transportation between the burgeoning urban centers it connected, the Boston & Worcester resisted any low-rate experiments to attract freight coming out of the Erie Canal. Until their official merger in 1868 (forming the Boston & Albany Railroad), the two companies negotiated endlessly and unsuccessfully to find a common strategy.[35]

These negotiations constituted a struggle for the soul of the Western Railroad. Roughly defined, it was a struggle between pragmatists and visionaries. Almost all the directors and major stockholders of the Boston & Worcester Railroad were pragmatists. Even Nathan Hale, editor of the *Boston Daily Advertiser* and so important earlier in promoting the vision of a railroad across Massachusetts, became more pragmatic as the first president of the Boston & Worcester. Over time, he became protective of its prerogatives and tightly focused on ensuring its financial success. He strongly opposed lowering his rail-road's freight rates to increase Boston's share of Erie Canal traffic. He success-fully blocked a merger with the Western Railroad in 1845, thereby delaying it twenty years. Other members of the Boston & Worcester board were even more parochial. William Jackson of Newton convinced the railroad to initiate a low-fare commuter service between his home town and Boston, while simultane-ously developing Newton's first major land subdivision. For its part, the West-ern Railroad directors were a mix of pragmatists and visionaries. The railroad's president for much of its early life was George Bliss, a shrewd, authoritarian pragmatist who saved the line from bankruptcy several times. When financial disaster threatened, he was not afraid to raise freight rates and passenger fares. Bliss was also leader of the Whig Party in the state legislature. He recognized how intertwined railroads and politics had become in the Bay State and that his railroad's financial difficulties were an easy target for the Democratic opposi-tion. (In fact, Democrat Marcus Morton narrowly defeated the Whig governor in 1842 on this issue.) Diametrically opposed to Bliss was Elias Hasket Derby III, leader of the visionary faction of the Western board. He was joined by Peter P. F. Degrand and Nathan Carruth. A Frenchman and stock broker, De-grand had helped found the Boston Stock Exchange and served as its first president. Although he would not have used the word, he was what one today would call a "lobbyist" for the state's railroad interests. From humble begin-nings, Nathan Carruth had made several fortunes, first in drugs, then in paint, and finally as an investor in the Old Colony Line. (The Old Colony Line served communities below Boston, along the so-called South Shore.)[36]

Derby had been an early, fervent booster of the Western Railroad. Born in Salem and a Harvard graduate, he studied law in the office of Daniel Webster before being admitted to the bar in 1829 and the Supreme Court two years later. He won distinction as a brilliant railroad lawyer and became president of the Old Colony Line, where he worked with Nathan Carruth. Derby wrote numerous articles about railroads for Boston newspapers, signing them sim-ply "Massachusetts." He was also a contributor to the *Atlantic Monthly* and the *Edinburgh Review*. Derby's articles and speeches blended past and future. They were steeped in nostalgia for Massachusetts' former maritime glory—in which his grandfather had played such a significant role—but also reported on the latest advances in railroad technology. Above all else, Derby was convinced the Western Railroad could restore his state's bygone commercial glory. If the railroad was properly managed, he predicted, the Erie Canal would "pour its tribute into our city." Derby was sure that low freight rates on the Western were key to diverting the canal's grain and flour away from New York City to Boston. Derby hated George Bliss for his lack of vision and high-rate policy. He hectored Bliss for every mistake the railroad made. Derby warned Bliss not to purchase the railroad's shoddy locomotives but to subject them to further testing. Derby also criticized the railroad's abysmal safety record. Similarly, he pointed out the flaws in the Western's facilities on the Hudson River and the inefficiency with which cargo was transferred from canal boats to the railroad. The greatest costs were incurred in "the *handling* of merchandise," Derby told Bliss. "I would as far as possible advocate the doctrine of *touch not, taste not, handle not,* if dividends are desirable and [Hudson River] packets are to be competed with." Derby was fast losing patience with the Western Railroad as a way to realize his dream of restoring Bay State commerce. He began looking for a better opportunity.[37]

In spite of a shaky start, Massachusetts railroads proliferated rapidly. By 1850, Boston became the most railroad-connected city in America. Fittingly, Bay State authors endeavored to describe the railroad's arrival in their communities. Whether they welcomed the railroad or not, they agreed there had never been anything like it. Whereas textile factories had had their rustic precursors in flour mills and cooperages, where men and machines worked together under the same roof—railroads were completely new, defied the imagination with their power and speed, and radically transformed everyday life. After retreating to Walden Pond to escape his neighbors, Henry David Thoreau had his repose violated by the whistle of a nearby locomotive "like the scream of a hawk" and seconds later by the "rattle of railroad cars." During his first train ride, Ralph Waldo Emerson noted the extreme distortion of his senses. "The very permanence of matter seems compromised," he writes, "[and] hitherto esteemed symbols of stability do absolutely dance by you." Herman Melville's friend and fellow novelist Nathaniel Hawthorne was similarly impressed and confounded. In *The House of Seven Gables,* Hawthorne describes Clifford Pyncheon and his sister Hepzibah's first railroad experience. They climb aboard in Concord, Massachusetts, but must make haste. The locomotive is "fretting and fuming like a steed, impatient for a headlong rush." With the clang of a bell, they are off. "Meanwhile, looking from the window, they could see the world race past them. At one moment, they were rattling through solitude; the next, a village had grown up around them; a few breaths more, and it had vanished, as if swallowed by an earthquake." Sensing the same instability Emerson had, Hawthorne writes "The spires of meeting houses seemed set adrift from their foundations; the broad-based hills glided away. Everything was unfixed from its agelong rest, and moving at whirlwind speed in a direction opposite to our own." Turning his attention to the inside of the train, Hawthorne gives us a uniformed conductor passing through a quiet group of passengers. Since it is 1851, conductors now wear uniforms. So, too, passengers are well-behaved. (Because railroads were America's first form of mass transportation, the public had to *learn* how to behave on them.) Some of the passengers have ridden trains before and now take them for granted. They are either reading or conversing with their neighbors, seemingly oblivious to "so much noisy strength at work in their behalf." Clifford Pyncheon is a daguerreotypist by profession, an early photographer in a society still wedded to painted portraiture. As a thoroughly modern man, he engages an older gentleman in a debate about the merits of riding on a train versus sitting at home by one's fireside. Clifford tells the old fellow "that this admirable invention of the railroad—with the vast and inevitable improvements to be looked for, both as to speed and convenience— is destined to do away with those stale ideas of home and fireside, and substitute something better." In his innocent riposte, the old fellow asks, "What can be better for a man than his own parlor and chimney corner?" This only sets off his young acquaintance on a lengthy diatribe against things past. Such a simple life "possessed a charm," Clifford concedes. But what about its drawbacks? There was "hunger and thirst, inclement weather, hot sunshine, and weary foot-blistering marches over barren and ugly tracks," Clifford reminds him. "These railroads," he declares, "are positively the greatest blessing that the ages have wrought out for us. They give us wings; they annihilate the toil and dust of pilgrimage; they spiritualize travel!" Up against Clifford's nearly maniacal belief in human perfection through technical progress, the old gentleman stands little hope of persuasion. Still, Hawthorne colors Clifford's argument with something nearly as unhinged and unstable as the landscape flying by outside the train. Finally, Clifford's sister Hepzibah tells her brother to control his enthusiasm and be quiet. "They think you mad."[38]

While exaggerating his characters' views to the point of parody, Hawthorne was tapping into a well-established strain of enthusiasm for modern technology among many of his countrymen and its outright rejection by others. Similar arguments had shadowed the debate over the Western Railroad and would reoccur for and against the Hoosac Tunnel. These debates had a history. Two decades earlier, American essayist Timothy Walker had published his "Defense of Mechanical Philosophy" in the *North American Review,* refuting British historian Thomas Carlyle's dire assessment of the mechanical age in the *Edinburgh Review.* Walker had graduated from Harvard in 1826 at the head of his class, taught school under the accomplished scholar George Bancroft, and prepared for a career in the law with Justice Joseph Story. The *North American Review* was the mouthpiece of Bay State intellectuals, many soon to join the Whig Party, who believed in the improvement of society through technical progress. Carlyle had argued that machines exercised a "pernicious sway" over men's minds, that an "inward sense of machinery" corrupted their thoughts. "Poetry," he observed, was "no longer without a scientific exposition." Philosophy had also been debased. "The whole doctrine of [John] Locke is mechanical," he declared. Dismantling Carlyle's arguments in a lawyerly fashion, Walker accuses Carlyle of mysticism for such beliefs. Modern man has become more thoughtful because of labor-saving machines, Walker argues. The ancient Greeks, lauded by Carlyle, had achieved their high level of cultivation because of slave labor. Possibly reflecting his Massachusetts origins, Walker describes the past as a struggle to survive upon a "bleak, naked, incommodious earth." He has no desire to trade the amenities of modern life for the "low mud-walled cottages" of an earlier age. Nor is Walker willing to give up the improvements to nature's rugged landscape made possible by mechanical innovations. "Where she [nature] has denied us rivers, Mechanism has supplied them," Walker declares. "Where her mountains have been found in the way, Mechanism has boldly levelled or cut through them. As if her earth were not good enough for wheels, Mechanism travels it upon iron rails." Like Hawthorne's fictional character Clifford, Walker enters the sublime in describing things mechanical. "Examine the endless variety of machinery which man has created. Mark how all the complicated movements co-operate in beautiful concert." Drawing on the Puritan notion that man and God's work are contiguous, Walker goes further. He *sanctifies* the mechanical arts: "We believe that there is nothing irreverent in the assertion that the finite mind in no respect approximates so nearly to a resemblance of the Infinite Mind, as in the subjugation of matter, through the aid of Mechanism, to fixed and beneficial laws—to laws ordained by God, but discovered and applied by man."[39]

Like it or not, Massachusetts was in the thrall of "mechanism" by the mid-nineteenth century. The build-out of railroads radiating from Boston and the scaling-up of textile mills in Waltham and Lowell signaled its arrival. Even the Old World was forced to take notice. During the 1850s, a British Parliamentary Commission visited Massachusetts to investigate its factory system and production techniques. After visiting both large and small manufacturing operations, these English investigators found that the average laborer there possessed a faith in technology and a talent for innovation largely absent in their own country. "From the habits of early life and the diffusion of knowledge by free schools," the investigators write, "there exists generally among the mechanics of New England a vivacity in inquiring into the principles of the science to which they are practically devoted." Compulsory education during part of each year, the investigators believe, "lays a foundation for the wide-spread intelligence which prevails amongst the factory operatives." These young workers were "daily improving upon the lessons learned from their older and more experienced compeers." Early support for universal education in Massachusetts, initially intended to school the young in the scriptures and prepare them for membership in the Puritan church, had spawned a spirit of improvement on the factory floor.[40]

There were thousands of poor young men and women working in factories across Massachusetts. It was not surprising that a few of them rose to wealth and prosperity during a period of rapid industrialization. One was Alvah Crocker, who began his working life in a Leominster paper mill at age eight. Through hard work and education, Crocker would become a powerful paper baron, found a bank, serve as president of a railroad, and represent his district in Congress. Although his upbringing differed markedly from that of Francis Cabot Lowell, Crocker was an industrial innovator like his predecessor and championed the economic advancement of his state much as Lowell had. Among his many distinctions, Crocker would be known as the "Father of the Hoosac Tunnel."[41]

3

MASSACHUSETTS LOOKS WEST

AS AMERICA EXPANDED to the West, that region increasingly fueled the nation's growth—every state, every part of every state, and every community sought westward avenues to bolster their prosperity. The lure of the West had been the inspiration for the Erie Canal and only intensified as the West became more productive and threw off wealth for those fortunate enough to access it. Alvah Crocker not only understood this; it became his lifelong obsession.

The Crockers lived in the village of Leominster, on the Nashua River in northernmost Worcester County. Though rumor had it the family came from Puritan stock and was once prosperous, the Crockers were now decidedly poor. Alvah was born in 1801, the first of seven boys. His childhood was not easy. His father, Deacon Samuel Crocker, had founded the local Baptist church and often held gospel meetings at home. He was a hard, uncompromising patriarch, who believed that godliness and hard work were synonymous. He sent his oldest son to work at Israel Nichol's paper mill, where Samuel himself was a foreman, at age eight. Alvah worked twelve-hour days for twelve cents a day. At the time, Massachusetts boasted more than forty paper mills. The high rate of literacy in the state and the invention of the rotary printing press had created an insatiable demand for paper. Paper was still made by hand from old rags (wood pulp would not be used until the 1840s) and was an artisanal trade. Alvah Crocker had mastered it by his teens.[1]

Even as a young boy, Alvah's intellectual curiosity took him beyond reading the scriptures. He borrowed books from Israel Nichols and others in Leominster who were impressed by his desire for learning. By sixteen, he had saved $60 and entered Groton Academy, determined to stay as long as his money lasted. In 1819, he wrote a friend that he had read seventy-eight lines of Virgil.

The following year, he abandoned his hopes of attending college and decided to cast his lot with the business he knew best. Crocker went to work for a paper mill in Franklin, New Hampshire, and after two years moved on to Fitchburg, Massachusetts, which was fast becoming the leading paper-making center in the United States. In 1826, Crocker borrowed enough money to start his own paper mill. He produced printing paper, writing paper, and wrapping paper. He gradually mechanized his paper-making process with drying cylinders and pulp makers. However, he faced a great barrier to success, a logistical one. Fitchburg was one of many towns in northern Massachusetts lacking a decent transportation connection to Boston. Crocker was often forced to deliver paper there by horse and wagon, sometimes in terrible weather. He never forgot the experience. He became a fervent advocate of internal improvements, speaking frequently at public gatherings. He joined the rising Whig Party and served in the state legislature between 1837 and 1843. Crocker made several trips to the nation's capital to seek a remission in the iron tariff and support for his railroad ideas. He also rode back and forth on the Baltimore & Ohio Railroad to study its design and operating methods. He thought it "the best railroad I was ever on." He met with newly elected President James Polk on Post Office appointments but wrote Polk off as a "second rate man." Crocker was delighted whenever he departed Washington, D.C., calling it a "degraded, hollow-hearted place, which I loathe and abhor."[2]

In the numerous photographs of him taken during his long life, Alvah Crocker's expression is always the same. His deep-set eyes are clear and intense below a strong brow and high forehead. Tightly clipped mutton chops and beard frame but do not obscure his powerfully set jaw and the perfectly straight mouth. His stare is intense and focused but reveals nothing of what he is thinking. Mostly through Crocker's efforts, the Fitchburg Railroad was chartered in 1842 and completed in 1845. It had capitalization of $1,322,500, funded entirely through the sale of stock, and it spanned forty-nine miles from Fitchburg to Charlestown, a deep-water port next to Boston. The railroad clearly addressed the needs of communities along its line and faithfully paid dividends of 8 to 10 percent per year. Three passenger trains and one freight train traveled the line daily. By 1844, Crocker had received a charter for the Vermont & Massachusetts Railroad, which, when completed in 1850, ran west from Fitchburg to Greenfield, Massachusetts, on the Connecticut River (thirty-five miles north of Springfield). Incorporating several existing tracks, the line then meandered north to Brattleboro and Rutland, Vermont, before turning south to Troy, New York, on the Hudson River. Though it was painfully circuitous in its route and inefficient for through traffic, Crocker had connected Boston and the Hudson River.[3]

One reason the Fitchburg Railroad was so well-financed had to do with the efforts of Elias Hasket Derby, who joined the railroad's board in 1842. While still involved in the Western Railroad, he did not seek reelection to its board once he joined forces with Crocker. Crocker and Derby made a congenial pair. They proved to be an effective fundraising team, the former scouring the countryside for modest purchases of Fitchburg Railroad stock from farmers and tradesmen and the latter soliciting Boston's business community for more substantial tranches. Their timing was perfect. Massachusetts and the nation were finally emerging from the long depression that had begun with the 1837 financial panic. The balance of the 1840s were boom times for Bay State manufacturing. Crocker and Derby even traveled to Europe together to study railroad construction techniques. While in England, they rode the Manchester & Leeds Railroad through the two-mile-long Summit Tunnel. They asked questions and made notes on the tunnel's construction method and cost. Clearly, the Fitchburg Railroad was just one part of a larger plan.[4]

Derby's account of his and Crocker's European tour, published in 1843 and titled *Two Months Abroad,* focuses mostly on England. One of the few things Derby admires there is its lush, fertile countryside, "superior in soil to any part of Massachusetts." However, the laborers in those fields strike him as "dull and ill fed." Factory workers, Derby observes, are even worse off, the majority being "degraded, illiterate, and depraved." The root of the problem is that England invests too little in educating its population. This offends Derby's Puritan sensibilities. "The factory operative with us is better fed and paid, and is more moral and intelligent," he adds. "Let her [England] look at Lowell," he continues, "in a State settled by Puritans, and see how the descendants of those Puritans treat their brother Christians." Nor is Derby impressed with England's system of railroads. The stations are too lavish and roadbeds, trestles, and roundhouses overbuilt. To fund such an extravagant infrastructure, passengers must pay too much. "The charges are so high that the lower orders are debarred from the railroads." The country's freight service is also underdeveloped and suffers from the same problem of overpricing, according to Derby. While his commitment to low pricing and prudence in capital spending had alienated Derby from the board of the Western Railroad, these sentiments resonated with his friend Alvah Crocker.[5]

By the time he and Derby met, Crocker had consolidated several paper mills in Fitchburg and nearby towns and was considered one of the Bay State's leading "paper barons." His merger with the highly successful but ailing paper manufacturer Gardner Burbank, to form Crocker, Burbank & Company, was an important step in this process. Crocker had channeled the religious intensity of his early upbringing into a driving work ethic. He soon opened a furniture factory, founded an insurance company, and started a bank. His portrait appeared on its banknotes. Crocker was now in a position to advocate even more aggressively for those communities and industries he felt were underserved by the state's transportation system. Much of his resentment was directed at the Western Railroad. In a state as regionally divided and parochial in its economic interests as Massachusetts, one railroad line connecting Boston to the Hudson River, especially one running along its southern flank, could never satisfy everyone. Such "balkanization" occurred in many states as railroad networks expanded and every community sought its own railroad connection. Still, much of this sentiment was justified by the poor service from major railroad lines like the Western to these communities. For example, merchants in North Adams, Massachusetts, complained bitterly about the Western's feeder line from Pittsfield to their community. As the Western sought to increase its profitability, it raised freight rates on these feeder lines and scrimped on maintenance. Though it had grown in both population and industry, Crocker believed his own town had also been neglected by the state's railroad network. More than ever, he was committed to bringing increased prosperity to "this tortuous winding gorge, this broken, rockbound Fitchburg." Crocker would not rest until the town had its own rail connection to the Hudson River and access to the immense bounty of the West beyond it.[6]

Derby was motivated by somewhat broader interests. He was much more attuned to the needs of the entire state and its rivalry with New York. That rivalry was for grain and other foodstuffs coming from the West to the eastern seaboard for export. During the second half of the1840s, railroad track mileage in the western grain-producing states tripled. These new tracks improved the access of farmers in these states to Great Lakes ports and the Erie Canal. Furthermore, the Irish potato famine beginning in 1845 created a surge in demand for American foodstuffs. As a result of these developments, the volume of grain moving down the Erie Canal rose dramatically. Whereas this volume of grain had been just over two hundred thousand tons in 1842, when the Western Railroad first opened, it exceeded almost ten times that amount a decade later. And, yet, as Derby and his fellow critics pointed out, the Western Railroad seemed incapable of garnering its fair share of this windfall. Its share of western grain continued to fall, with Boston steadily losing out to New York as the port of export for these goods. In truth, most shareholders of the Western Railroad had little interest in hauling bulky, unprofitable freight across the state. By the late 1840s, the railroad had aggressively increased its freight rates and passenger fares and was returning an 8 percent dividend to its shareholders. Like its sister company, the Boston & Worcester, the Western Railroad had given up the dream of diverting Erie Canal traffic from New York to Boston. In Derby's view, Crocker's plans for a railroad along the northern tier of Massachusetts offered a second chance for the state to access the cornucopia of riches pouring out of the Great West.[7]

When Crocker and Derby petitioned the Massachusetts legislature for the Troy & Greenfield Railroad's state charter in 1848—with access to the Hudson River via a tunnel through the Hoosac Mountain—local citizens and commercial interests split north and south for and against the project. Naturally, representatives of the Western Railroad opposed the idea of a competing railroad across the state. The cities of Worcester, Springfield, and Pittsfield, all served by the Western Railroad, wielded considerable power in the Massachusetts legislature, and many of their representatives were substantial Western stockholders. Nonetheless, these oppositional forces underestimated the degree of support among northern towns for the new railroad. Crocker and Derby were well-organized and held public meetings and formed political caucuses all along the proposed line. A meeting in North Adams, where the west portal of the Hoosac Tunnel would be located, was typical. On November 3, 1847, Alvah Crocker presided over a packed crowd in the local Baptist Church. A. F. Edwards, who had surveyed the Troy & Greenfield's route and would serve as its first chief engineer, described the planned construction of the railroad through the community. A local booster named Roger Leavitt was ecstatic about its potential: "Gentlemen, Nature planned the valleys of Deerfield and the Hoosac and left this bluff to test the perseverance of man. Someday this route will become the great thoroughfare from Liverpool to Pekin [sic]." Men like Leavitt possessed a surprisingly global vision and needed little urging from Crocker to ignite their community's grand ambitions. They instructed their legislators to deliver that vision to Boston. In response to those who claimed a second cross-state railroad was unnecessary, local newspapers heralded the new railroad as transformational for the region. In answer to those who said the Hoosac Mountain could not be breached, Crocker and Derby rolled out their secret weapon: Edward Hitchcock. Besides being president of Amherst College, Hitchcock was also the state geologist. More than anyone else in America, he had popularized geology and made it a part of the nation's academic curriculum. First published in 1840, his *Elementary Geology* would go through thirty-one printings before the Civil War. Massachusetts legislators listened intently as Hitchcock explained what miners could expect to find inside the Hoosac Mountain. The mountain was an "extremely simple formation," composed almost entirely of mica slate with nodules of quartz here and there. "There is no question but that the rock is easy for drilling," Hitchcock observed. What is more, the solid consistency of the mountain meant that the tunnel would not require any internal structure to support it. "I do not believe there will be any masonry or arching required," Hitchcock explained. Nor would internal flooding be a problem. The tunnel would "go below where water percolates" and, he predicted, "will be found to be dry." Given these assurances from such an esteemed expert—and the fact that the petitioners were not asking for state aid—the legislature granted a charter to the Troy & Greenfield Railroad, to include a tunnel through the Hoosac Mountain. It was a terrible mistake.[8]

As it turned out, obtaining a charter for the Troy & Greenfield Railroad was the easy part. Finding investors in the new line proved more difficult. Despite their talents for salesmanship, Crocker and Derby, now joined by Peter P. F. Degrand, were unable to convince the citizens of western Massachusetts to part with their hard-earned money for such a speculative venture. At this early stage, local enthusiasm for the Troy & Greenfield Railroad did not translate to financial support. With only a few thousand dollars in its treasury, the Troy & Greenfield Railroad was forced back to the state to ask for financial aid. The railroad requested a loan of $2 million, payable over twenty years at 5 percent and secured by a mortgage on the assets of the line. This time, however, the representatives of the Western Railroad were better prepared. They attacked the Troy & Greenfield full bore in newspapers, pamphlets, and on the floor of the legislature. While advocates of the new railroad would ultimately prevail, supporters of the Western Railroad were able to delay state aid for three years. The first round in this legislative marathon took place during the spring of 1851.[9]

A series of articles in the *Boston Daily Advertiser,* the mouthpiece for the Western Railroad, argued that tunneling through the Hoosac Mountain would be an unprecedented, "Herculean" task, costing far in excess of the funds being requested by the petitioners and taking much longer than anyone could reasonably predict. Loammi Baldwin's earlier study of the mountain had been conducted when "engineering was in its infancy," the newspaper reminded its readers. Regarding Edward Hitchcock's more recent analysis, the paper cautioned that Hitchcock was still "mortal and fallible" in spite of his many accomplishments. "His eye of science cannot penetrate the fifteen hundred feet into the bosom of this mountain, and tell us with certainty and accuracy, what material lies imbedded there; and none but the eye of the Almighty can." More objective surveys of the Hoosac Mountain, the newspaper continued, showed that it contained a variety of problematic materials unfavorable to tunnel boring—from gneiss, "the hardest kind of rock," to soft, pliable soprolite, subject to "disintegration when exposed to air and requiring roofing." Overcoming these problems could cost multiples of what the petitioners were asking for and require "patience equal to that of Job" to complete the tunnel. One member of the Massachusetts Senate characterized the entire project "a wild one, baseless and visionary as a dream of childhood."[10]

Opponents of the Troy & Greenfield Railroad also argued that the new line was unnecessary. They pointed out that feeder lines had been extended from the Western Railroad to many communities in northern Massachusetts. "There is scarcely a valley north of the Western Railroad, where a railroad is not built." Drawing on the homespun logic of his rural constituency, one legislator tried to simplify the issue: "We don't call that man a good, frugal, practical farmer, who buys and feeds two yoke of oxen to subdue his soil and do his work, when it can be as well or better done with one yoke." Nor, the speaker continued, do we call him "a wise manager who divides his capital between two shops in the same village to rival and ruin each other." A second railroad across Massachusetts would be redundant and threaten the health of the Western Railroad. "One good road, well sustained and well managed, is better for the public than two sickly, pining, half sustained roads."[11]

But was the Western Railroad well managed? While on the Western's board and now as an advocate for the Troy & Greenfield Railroad, Derby had argued that it was not, mainly because it had failed to divert Erie Canal grain from New York to Boston. After all, this had been the reason for building the Western Railroad in the first place. In responding to this accusation, representatives of the Western Railroad chose a surprising, and potentially dangerous, line of defense. They argued that no railroad across Massachusetts could compete with shipping by water down the Hudson River, neither the Western Railroad nor any new competing one. In making this argument, opponents of the Troy & Greenfield not only admitted to the failure of the Western Railroad to capture Erie Canal freight but went into considerable detail about it. During the past year, they explained, the Western had carried only 60,900 tons of the two million or so tons of grain that had arrived at Albany over the canal. This was because it cost 28 cents to transport a barrel of grain on the Western Railroad versus 12 cents to ship it down the Hudson River by water. In other words, the opponents of the new railroad were saying, competing for traffic coming out of the Erie Canal was a lost cause, an economic impossibility for any Massachusetts railroad. As a Western Railroad pamphlet flatly stated, "No railroad to Boston can compete with the Hudson River."[12]

Making such an argument was a tactical mistake, and the "tunnelites"— as they were now called—were quick to take advantage of it. Opponents of the Troy & Greenfield were asking the Massachusetts public to abandon its dream of capturing a fair share of the western bounty coming over the Erie Canal and thereby restoring Boston's commercial greatness. During a period of shifting trade flows and startling technical innovations, such fixed and dismal logic was impossible for many to accept. "Shall we shrink from such a work as this," tunnelite Whiting Griswold asked his legislative colleagues, "an enduring, stupendous monument of the indomitable energy, of the industrial, intellectual, social and moral progress?" The Erie Canal was "a great work of internal improvement [which] formed a new era in the history of that State." Even now, he noted, New York was building a railroad from the Erie Canal to New York City along the very banks of the Hudson River. If railroad transport was so uncompetitive with shipping by water, why would New York do such a thing? The answer, Griswold declared, was that railroads were a God-given invention to harness the forces of nature and channel its bounty in ways heretofore unimaginable. Railroads were "the very instruments designed by Providence to develop the incalculable resources of the earth; to divert the minds of men from the arts of war to those of peace; foretold in the Holy Writ when the sword shall be beaten into the ploughshare, and the spear into the pruning hook." Why would Massachusetts deny itself the blessings of this marvelous invention? Why would it "drive this immense business between West and East around through other States, instead of opening the sides of this mountain and suffering it to pass through our State?" Griswold was setting forth the two themes that would dominate tunnelite rhetoric for years to come: first, the lure of the Great West, with its promise of unlimited wealth, and, second, the power of human invention to access that wealth for the Bay State. These powerful metaphors, evoking widely accepted notions of western agricultural abundance and the transforming power of innovative technology, would do battle with realism and practicality over what to do with the Hoosac Mountain.[13]

While the tunnelites were more passionate than the Western's defenders of the status quo, both sides of the Hoosac Tunnel debate marshaled reams of data and paraded a variety of expert witnesses before the legislature to bolster their arguments. Defenders of the Western Railroad focused mostly on cost information from tunnel projects at home and abroad, citing the price per linear foot of core material removed from the Kingwood, Rockdale, Woodhead, and other recently completed tunnels. This information was used to estimate the overall cost and duration of work on the Hoosac Tunnel. The type of material encountered, the number of vertical shafts drilled, manpower employed, need for arching, and other specifics were cited in each case. In response to this data, the tunnelites fell back on Edward Hitchcock's analysis and their belief that the technical advances being made in tunnel engineering would allow the Hoosac Mountain to be easily breached. The tunnelites tended to skim over present concerns and focus on the future possibilities. They emphasized the longer-term advantages their new cross-state railroad would have over the existing one in capturing the sought-after Erie Canal trade. These advantages would include shorter distances, lower grades, and gentler curves. The tunnelites frequently resorted to pseudo-scientific formulations to equate more favorable curves and grades on their proposed road with shorter mileage versus the Western. They also cited the tremendous debt burden the Western's difficult construction had saddled it with. In short, the Troy & Greenfield Railroad would be shorter, straighter, flatter, and nearly debt-free. This would allow it to compete on price with Hudson River freight, which the Western Railroad could never do.[14]

The tunnelites won the first round of their contest for state aid when the Massachusetts Senate approved their bill in April 1851. It was then passed to the House, where it was debated for two more months but ultimately defeated by a vote of 227 to 108. During the balance of 1851 and through the next year, the tunnelites regrouped and prepared for their next assault on the State House. There were a number of reasons why the Hoosac Tunnel debate was more contentious and protracted than the Western Railroad had been two decades earlier. Of course, the tunnel was a more challenging undertaking and represented a second railroad across the state. No one could predict the eventual cost of the tunnel, and the Western tenaciously defended its cross-state monopoly. But there were other reasons as well. Both the economic and political environments were less favorable for the Troy & Greenfield than they had been for the Western twenty years before. To begin with, the state's manufacturing elite, the Boston Associates and similar firms, were no longer as prosperous and powerful as they had once been. The proliferation of new textile mills and increased out-of-sate competition had led to overcapacity, saturated markets, and stagnating profits. Boston's wealthiest investors had grown more risk averse, preferring the guaranteed returns from banks and insurance companies over speculative infrastructure schemes. Although it had supported the Western Railroad generously, little private investment for the Hoosac Tunnel would come from the state's moneyed class. This wealthy elite had also watched its political clout erode, as the Whig Party collapsed under the pressure of national events. At the center of these events was the issue of slavery.[15]

In fact, the once unified and unassailable Massachusetts Whig Party—ideologically committed to internal improvements of almost any kind—no longer existed by the early 1850s. The controversy over the Mexican War, fought between 1846 and 1848, and the fallout from the Compromise of 1850, contentious legislation meant to absorb the territorial spoils of that war while appeasing the southern slavocracy, fractured the Whig Party nationally and in the various states. In Massachusetts, the party was hopelessly divided between "Cotton Whigs," representing the state's manufacturing interests and supportive of the Compromise of 1850, and the "Conscience Whigs," antislavery reformers vehemently opposed to the Compromise of 1850 and, in particular, the so-called Fugitive Slave Law embodied in it. (This law required escaped slaves living in the North to be returned by federal marshals to their southern masters.) While both groups laid claim to the Puritan legacy of enterprise and morality, that unique duality had been badly damaged along with the communitarian unity that had given it such strength. In short, post-Puritan solidarity collapsed before the slavery crisis. After the crushing defeat of its presidential candidate, Winfield Scott, in 1852, the Whig Party shambled off the national political stage never to return. In Massachusetts, the former heartland of the Whig Party, Democrats and the nascent Free-Soil movement came together to capitalize on Whig disarray. A coalition of Democrats and Free-Soilers took control of the Massachusetts legislature in 1851 and 1852. While the coalition was fragile and short-lived, its members shared an unfavorable view of state aid to railroads and were responsible for defeating the Troy & Greenfield's $2 million loan bill. Only when a feeble, cobbled-together vestige of the Whig Party recaptured the state legislature in 1853, their last gasp as a party in Massachusetts, were the tunnelites in a position to relaunch their appeal for state aid. It was a brief window of opportunity before state politics deteriorated further.[16]

After so long a struggle, both sides of the tunnel debate were tired and bitter. When the Troy & Greenfield approached the legislature in February 1853 with yet another petition for the $2 million loan, Ansel Phelps, attorney for the Western Railroad, was dismissive. He called the new petition a "newly vamped-up" version of what had been presented before. "It has an ancient, fish-like smell," he said. Elias Hasket Derby was on his feet immediately. "We are met by the Western Railroad as we have been met before—with sneers," he countered. "Sneers are not arguments to convince. Fulton and his steam boat and George Stevenson with his railroad engine were also met with sneers." The session was off to a bad start. Derby once more accused the Western Railroad of failing to fulfill its primary obligation to the people of Massachusetts. "It has not turned the mighty trade of the West to Boston," he reiterated. Not only had Massachusetts fallen behind Pennsylvania, Maryland, and Virginia in its share of western trade, but the Western Railroad was carrying less Erie Canal grain over its tracks every year. Furthermore, none of these other states had let tunneling stand in the way of their ambitions. The Baltimore & Ohio Railroad now had fifteen tunnels, including the magnificent Kingwood Tunnel, Derby explained. The Pennsylvania Railroad was replacing its inclined planes with tunnels. Derby noted that there was an "abundance of precedence in tunneling" and promised the Troy & Greenfield would avail itself of this experience. Already, the company had the "Excavator," a machine capable of boring two feet per hour, waiting at the base of the Hoosac Mountain. "Should it prove successful, of which there is scarcely any doubt, the cost of the tunnel will be reduced to $1,000,000." He invited members of the legislature to witness this mechanical marvel for themselves.[17]

Who wouldn't have gone to see such a technological wonder? On March 15, 1853, during a break in the State House calendar, several dozen legislators travelled on the Fitchburg Railroad out to Greenfield, Massachusetts. Located on the west bank of the Connecticut River, just north of where the Deerfield River joins it, the town was known for its cutlery and tool-making industries. Its three thousand citizens were abuzz with speculation about the new railroad coming through their town. The next day, the delegation went by horse and wagon up the Deerfield Valley to Rice's Hotel, less than a mile from the intended east portal of the tunnel. There they met representatives from North Adams, where the tunnel's west portal was being laid out, and from Troy, New York, where the completed railroad would eventually reach the Hudson. Even though the weather was frigid, the Deerfield Valley was breathtakingly beautiful. It was one of the few fertile areas in the state and had been home to the Algonquin-speaking Pocumtuck nation before European settlement. The Pocumtuck had given the Hoosac Mountain its name, "place of stones." Amongst the valley's bucolic splendor, the "Excavator" must have seemed shockingly out of place. Built of cast iron and weighing almost a hundred tons, its twenty-five-foot-wide cutting wheel pressed tight against the Hoosac Mountain. Its proper name was "Wilson's Patented Stone-Cutting Machine," and it had taken nine months to cart the machine piece-by-piece up the valley and assemble its cogs, shafts, wheels, and bolts. It had been built by Richards, Munn & Company of South Boston at a cost of $22,000. When the signal was given, the great machine began smoking and lurched into the mountain. As iron met rock, the ground shook and the scream was deafening. After fifteen minutes, this violation of the senses and the natural order ceased completely. The chief engineer, A. F. Edwards, was concerned that the freezing temperatures could damage the machine if it ran any longer. Still, it had ground four and a quarter inches into the mountain. While a small first step, the great bore had begun![18]

The tour was not over for the legislators. They rode by sleigh to the top of the Hoosac Mountain. They visited the town of Florida, some thousand feet or more above where the tunnel would pass. At Troy, New York, the delegation inspected the bridge over the Hudson River and the foundation of an immense terminal intended to handle freight to and from Boston. From there, the legislators traveled down to New York City, where they visited the Empire Stone-Cutting Company. Clasping their ears, they witnessed machines cutting Connecticut red sandstone, blocks of "greywacke," which was harder than Quincy granite, they were told, and, importantly, samples of mica slate brought down from the Hoosac Mountain. They were informed that these experiments had replicated what they were unable to see at the mountain due to freezing weather. "The principle of the machine at the tunnel and those operated at the company are identical," their report stated. The delegation returned to Boston satisfied with the results of their investigation.[19]

When the debate over the $2 million loan recommenced later in March, the Massachusetts legislature seemed more sympathetic to tunnelite optimism regarding technical innovation and its impact on the tunnel's feasibility. Derby and Crocker pushed their advantage. Tunnel construction was no longer experimental, they argued. Money was cheap, below 5 percent. Erie Canal freight was burgeoning, but Massachusetts was getting little of it. "Shall the old Bay State be thrown behind in this race?" Derby asked. He recalled the state's glorious maritime past and envisioned its imminent revival: "When the [Troy & Greenfield] line is completed, you will see at one end of the tunnel an enormous business, and at the other end a noble harbor, the finest in the world. You will see a metropolis full of enterprise, intelligence and industry, and a population seeking employment in navigation." Derby called up a vision of Boston Harbor crowded with grain-laden hulls bound for Europe and wharves teeming with activity not seen since the halcyon days of Massachusetts shipping. It was as if Derby's grandfather was speaking through him.[20]

The parade of expert witnesses continued. The tunnel's chief engineer, A. F. Edwards, presented new cost figures. Removal of core material would cost between $60 and $64 per linear foot. At 25,000 feet in length, or four and three-quarter miles, the cost of the tunnel would not exceed $1,225,000. However, he had adjusted his overall cost estimate to $1,675,432 to cover contingencies. James Hayward, a civil engineer and mathematics instructor at Harvard, reported on his visits to the Marseilles Tunnel in France and the Woodhead Tunnel in England the year before. His findings bolstered Edwards's cost estimate. So, too, did the testimony of H. G. Knight, a machinist whose stone cutter was being used on the Pennsylvania tunnels. Under the onslaught of these witnesses and their dizzying array of facts, the Western Railroad's chief attorney, Ansel Phelps, became ill. Too sick to leave his bed, he asked for a delay. Derby and Crocker refused and ramped up their attack.[21]

Derby and Crocker were exhausted too. Their colleague, J. M. Keith, stepped in to close the case. He had waited six weeks for the Western's memorial and was out of patience, he told the legislature. Having obtained a copy of it from the printer, however, he set about demolishing it point-by-point. Regarding Western claims that the stone cutter at the tunnel would never work, those who had been there and seen it in operation knew this was false. Moreover, the Western Railroad had inflated the cost of tunnel construction and time it would take to complete it well beyond what expert witnesses had testified was reasonable. As for the Western's own accomplishments boasted of in its memorial, they were grossly exaggerated. In truth, the Western Railroad had "squandered the State's fortunes." It was no better than "a helpless infant nursed by the State." "Gentlemen," Keith concluded, "I have done with the memorial of the Western Railroad. I am glad to free myself from the mist and sophistries with which it would shroud the case." The argument before the state was crystal clear. "The State has stretched out her net-work of railroads to all parts of her territory, but she is still hemmed in for all practical purposes. Will she be content to remain thus circumscribed and shut out from the boundless and ever increasing trade of the West?" Keith asked. "Shall a small mountain barrier of four and a half miles in extent, paralyze her energies, while her sister States are perforating the Alleghenies in every direction?"[22]

The fear of being "hemmed in" as a small northeastern state—of being "shut out" from the broader dynamics of national development—was a powerful sentiment deployed effectively by Keith and the tunnelites. And why wouldn't Massachusetts fear being relegated to provincial inferiority? During the previous decade, the land mass of the United States had nearly doubled. In spite of Alvah Crocker's low opinion of him, President Polk had redrawn the nation's boundaries to include Oregon, Texas, California, and other western lands. (Most of this resulted from the Mexican War.) In fact, 1853 marked the high water mark of the country's territorial expansion before the Civil War. The discovery of gold in California and the expansion of the southern plantation system into Texas had stimulated the rapid settlement of this immense area. The sale of cheap government land and extension of railroad lines also contributed to this western migration of population. Not only did the center of America's population shift further away from Massachusetts, but the nation's total population increased by over half during this period. Famine in Ireland and political turmoil in Germany sent waves of immigrants to America. While some settled in urban centers like Boston, many joined the movement west. The spread of the telegraph after 1848 made Americans much more aware of their territorial immensity and burgeoning population. Massachusetts citizens were reminded of their dwarfed geographic status and reduced economic importance every time they opened their morning newspapers. Articles under the heading "Via Telegraph" reported on events at the forefront of the nation's ever-expanding frontier.[23]

These changes along with economic development and transportation improvements altered the relationship of America's great regions and the trade flows between them. Ten years earlier there had only been the East and the South—the former a traditionally commercial but newly industrialized region and the latter a staid but expanding cotton-based agricultural region. Now, there were three great regions—the highly industrialized East, the cotton South, and the fertile granary of the West. With improved transportation between the East and West, the exchange of manufactured products and foodstuffs between them had intensified and their relationship had grown closer. It began to overshadow the East's relationship with the South. The onset of the Crimean War in 1853 drove the flow of western grain to Atlantic ports to its highest level ever. Elias Hasket Derby understood these dynamics better than most and brought them into the debate before the Massachusetts legislature: "It was well remarked by Lord Bacon that three things constitute the greatness of the State—fertile lands, busy workshops, and an easy means of communication. We have in Massachusetts the workshops, the West beyond our mountains has the fertile fields, and what we need to constitute greatness, is an easy communication from those fields to our workshops."[24]

Derby finished by doing what he did best. He played upon his state's pride and indomitable spirit in overcoming past adversities. "It would seem that Providence has interposed this barrier to try the mettle of Massachusetts," he suggested. "I stood on the field of Bennington, in the valley of the Hoosac, where our fathers drew the sword and shed blood to protect their homes and fields," Derby recalled. "Our field of battle is the field of science," he continued. Massachusetts had "taken the lead in fisheries and in navigation, in every ocean; she has distinguished herself in the mechanical arts. Her manufactures are to her a mine of wealth. Must she now, when everything is so bright and full of promise before her, be bound down to a narrow and ruinous policy by fiat of the Western Railroad? God forbid!" The idea that Massachusetts had repeatedly reinvented itself and could again was imbedded in many of those Derby addressed. It was a powerful notion that they wanted reaffirmed.[25]

At the beginning of April, Ansel Phelps returned to the State House. He was pale and weak. He knew the Western Railroad had probably lost its case and the Troy & Greenfield Railroad would likely get its loan. He recounted a terrifying nightmare he had had while confined to his sick bed. "I have seen men toiling in the mountains for long years ahead," he said. He added that he "would advise a marble slab be placed at each opening to the tunnel, and on it should be inscribed 'A Monument to the Folly of Massachusetts.'" As the tunnelites began celebrating their anticipated victory, few paid much attention to his words.[26]

In May 1853, the Massachusetts House approved the $2 million loan to the Troy & Greenfield Railroad by a vote of 143 to 96. Given various amendments to it, however, the tunnel aid bill had to be returned to the Senate, where, after a series of unorthodox parliamentary maneuvers orchestrated by Western Railroad lobbyists, it was narrowly defeated. The tunnelites were incredulous, and the Senate president, Judge Charles Warren, was burned in effigy in North Adams. Adding insult to injury, the Western Railroad also brought its influence to bear at the Massachusetts Constitutional Convention, being held that same time in a separate venue. There, Western lobbyists tried to add an amendment to the new constitution being considered that would forbid future state loans to railroads like the Troy & Greenfield. Derby, Crocker, and Peter Degrand redoubled their efforts, not only to promote their own cause more forcefully in the legislature, but also to shame the Western Railroad for its questionable tactics and expensive lobbying efforts. A subsequent investigation castigated the Western Railroad for its behavior, especially for having used "public money" to oppose the tunnelites. (The Western Railroad was partially owned by the state, whereas the Troy & Greenfield was private.) Having clearly overplayed its hand, the Western Railroad was widely condemned, with sympathy now flowing to the tunnelites. In February 1854, the House approved the tunnel aid bill by 169 to 118 and the Senate by 16 to 7. Though a native of Worcester, Governor Emory Washburn reluctantly signed the tunnel bill into law. Known as the Railroad Assistance Act of 1854, it was the product of three years of intense political combat.[27]

The tunnelites had won the battle for state aid, but skepticism and concern continued to shadow the project from many quarters. The *Boston Post* criticized Governor Washburn for signing the tunnel bill and predicted that the project would cost far more than the $2 million loan amount. Similarly, the *Boston Advertiser* asserted that "the tunnel will never be completed [and] will be abandoned before it is finished." Furthermore, the conditions attached to the loan bill were highly restrictive. The state held a first mortgage on the assets of the railroad and would only release the loan in $100,000 installments, for every thousand feet of tunnel and seven miles of trackage completed to and from the tunnel. Ten percent of each installment had to be put into a sinking fund for the bonds behind the loan. Furthermore, the railroad had to raise $600,000 in stock, with 20 percent paid for in cash, before the first loan installment would be released. In short, the tunnelites had walked out of the State House with nothing in their pockets to restart work.[28]

In desperation, the Troy & Greenfield Railroad went behind the state's back and floated a $900,000 bond privately, using the same assets pledged to the state as collateral. While the language in the bond issuance acknowledged that the state had a first mortgage on the company's assets, such a bond itself was on its face a violation of Massachusetts statutes. However, nobody seemed to take notice. Nor did anyone ask what had happened to the "Excavator" in the year since legislators had visited the tunnel. Mysteriously, miners there continued to use the traditional hand-drill and black powder approach. It was slow, dangerous work that the Excavator was meant to obviate. At some point during the previous year, a full trial of this promising aid to tunneling had been carried out but failed completely. After boring ten feet into the Hoosac Mountain, the Excavator had seized up and frozen in place. Since it was locked in the mountain's firm embrace and could not be removed, a shed was constructed around it and a sign reading "Machine Shop" hung on it. The failed contraption inside was later scavenged for scrap. Fortunately, the Excavator's trial had been conducted nearby but not at the east portal itself. As a consequence, work commenced on the tunnel albeit using traditional drilling techniques. It was clear to those at the construction site that piercing the Hoosac Mountain was going to be more difficult than anticipated. Had the magnitude of this failure been known to those opposing the tunnel, it is doubtful the $2 million loan would have been approved.[29]

The problems of construction and financing took their toll on the Hoosac Tunnel's chief engineers and contractors. A. F. Edwards abandoned the project shortly after the failure of the Excavator. Likewise, Edward Serrell of New York signed on to the tunnel project with great fanfare but quickly ran up the tunnel's debt with little progress to show for it. Not only was the railroad continually short of funds, but serious construction issues surfaced at both ends of the tunnel. Working from the east portal, crews drilled and blasted through slate and mica with relative ease but were stopped dead by large veins of gneiss and quartz. Inside the west portal, miners encountered the opposite problem. There, soprolite, or rock that had degenerated into a soil-like consistency, proved unmanageable for the miners. Some called it "porridge stone" or "demoralized rock" because it flowed like mud when exposed to air. By this time, Edward Hitchcock's original analysis of the mountain was beginning to look uninformed and naïve. Neither Hitchcock nor anyone else really knew what geological surprises lay within the ancient core of the Hoosac Mountain.[30]

And, yet, these setbacks seemed temporary and superficial in light of Massachusetts' historical ability to innovate in times of crisis and repeatedly rejuvenate its economy. A letter from a sea captain, read during the final session of the tunnel debate, captured these sentiments: "I would ask who builds the best, biggest and handsomest ships in the world? Massachusetts,—Where do many go when finished? To New York." If Massachusetts wanted to keep its ships and ship captains at home, the sea captain continued, "It would be better for Massachusetts to give the two million dollars asked for, than not have the Hoosac Tunnel completed." The sea captain was appalled that any Massachusetts man would oppose a project so important to the state's future. Other voices also questioned why the state couldn't innovate once again and penetrate the mountain to access the western trade. An article in the *Boston Traveler* during the debate expressed this same optimism: "If improvements have been made in weaving and spinning and navigation, why may they not be in drilling?" In short, the steady advance of technology would soon overcome any problems at the mountain. Once the mountain was breached, the bounty of the West would flow into Boston. All that was needed now was the right man to do the job.[31]

Engraved portrait of John Winthrop, drawn by J. Chorley from an original picture by J. R. Penniman. Winthrop led the Puritans to Massachusetts during the 1630s and declared their colony a "City upon a Hill." The Puritans found the land difficult and soon turned to the sea to feed themselves. *Courtesy of the Massachusetts Historical Society, Boston, MA.*

Portrait of Elias Hasket Derby by James Frothingham, ca. 1800. Derby was one of the Massachusetts Bay Colony's great maritime innovators. Privateering made him America's first millionaire. He later developed Salem's lucrative China trade. *Courtesy of the Peabody Essex Museum, Salem, MA.*

Silhouette of Francis Cabot Lowell, ca. 1810, artist unknown. Another great Massachusetts innovator, Lowell transformed the state from merchant shipping to textile manufacturing. In spite of his immense wealth, he never had his portrait painted. *Courtesy of the Massachusetts Historical Society.*

This view of the Berkshire Mountains looks east from Herman Melville's writing desk. The language of his 1851 novel *Moby-Dick* indicates he knew work had begun on the Hoosac Tunnel. *Photo taken by Erin Hunt, used courtesy of the Arrowhead Melville Museum, Pittsfield, MA.*

Engraving by L. S. Punderson of Edward Hitchcock, president of Amherst College and America's reigning expert in geology. Hitchcock testified that the Hoosac Tunnel would be "easy for drilling" and "found to be dry." He was wrong on both counts. *Courtesy of the Massachusetts Historical Society.*

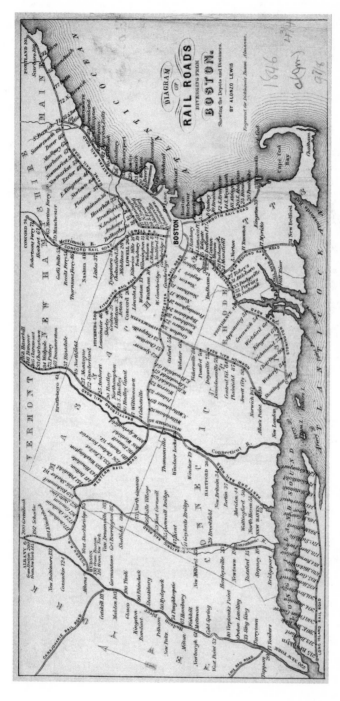

This 1846 railroad map of Massachusetts by Alonzo Lewis shows the Western Railroad and its sister line, the Worcester Railroad, running from Boston to the Hudson River. The two railroads could not cooperate to deliver western grain to Boston. The Fitchburg Railroad, heading northwest out of the city, would serve as the first section of the eventual Hoosac line along the state's northern tier to the Hudson. The Western Railroad would oppose the building of this competing railroad. *Courtesy of the Norman B. Leventhal Map Center at the Boston Public Library, Boston, MA*

Portrait of Alvah Crocker from *The Life and Times of Alvah Crocker*, by William Bond Wheelwright. Crocker went to work in a paper factory at age eight. Through study and hard work, he became a paper magnate in Fitchburg, Massachusetts, and founded the Fitchburg Railroad. He championed a railroad line across the northern tier of the state through the Hoosac Mountain. He would become known as the "father" of the Hoosac Tunnel. *Courtesy of the Massachusetts Historical Society.*

Rollstone Bank $100 note, featuring Crocker's portrait. In addition to his paper mills and railroad interests, Crocker became president of the Rollstone Bank in Fitchburg. Crocker was an excellent politician, representing his district in the state and national legislatures. *Courtesy of the Massachusetts Historical Society.*

Francis William Bird, illustration from an 1897 anonymous biographical sketch. Son of a prosperous paper merchant, Bird attended private schools and Brown College. There he studied moral philosophy and took up reformist causes such as abolition. He was a brilliant businessman, highly political, but always eccentric. *Courtesy of the Massachusetts Historical Society.*

Bird became a king-maker in Massachusetts politics and head of the "Bird Club." He was a Radical Republican and archenemy of the Hoosac Tunnel. He authored numerous pamphlets attacking the tunnel and its supporters. *Courtesy of the Massachusetts Historical Society.*

Committee's report of Progress, of **HOOSAC BORE** after an expenditure of $ 2,000,000.

This ca. 1856 broadside mocked the Hoosac Tunnel after it received a $2 million loan from the state of Massachusetts. The face of Alvah Crocker sits atop the locomotive about to enter the tunnel. This broadside may have given rise to the term "Great Bore." *Courtesy of the Massachusetts Historical Society.*

Herman Haupt at age thirty-four, shortly before he took on the Hoosac Tunnel. He became chief engineer in 1856 and struggled for six years, plagued by lack of funds, inadequate technology, and vilification by Francis Bird. Haupt would go on to serve as "Lincoln's railroad man" during the Civil War. However, Haupt's time at the Hoosac Tunnel ruined him financially. *Courtesy of the Yale University Libraries, New Haven, CT.*

Thomas Doane and his crew in front of the engineer's office at Hoosac Tunnel's east end, ca. 1863. Following Haupt's departure, the state of Massachusetts assumed ownership of the Hoosac Tunnel. Doane, seen here in a white smock, became chief engineer. Doane would preside over important technical advances, such as the Burleigh pneumatic drill, trinitroglycerin, and electrically detonated fuses. *Courtesy of Special Collections, Massachusetts Statehouse Library, Boston, MA.*

Miners preparing to descend into the west shaft. They are dressed for the wet, cold, often flooded conditions there. Like the Burleigh drill, their Otis safety elevator was an important Hoosac Tunnel invention. The miner on the right is holding a Burleigh drill. *Courtesy of Special Collections, Massachusetts Statehouse Library.*

On October 20, 1867, an explosion in the tunnel's central shaft building caused burning timbers and heavy equipment to fall into the shaft, killing thirteen miners trapped there. The next morning, a miner named Thomas Mallory, standing in left foreground, was lowered into the shaft and confirmed that his colleagues were lost. *Courtesy of Special Collections, Massachusetts Statehouse Library.*

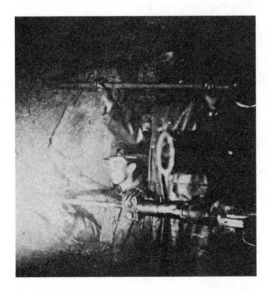

A rare photograph of miners deploying the Burleigh drill inside the Hoosac Mountain. *Courtesy of Special Collections, Massachusetts Statehouse Library.*

In this view of the east portal, ca. 1870, one of Thomas Doane's surveying huts with its conical roof and sighting pole stands next to the railroad tracks. The large cavity to the left is where Wilson's giant boring machine, the "Excavator," failed and was abandoned in 1853. A grocery store, school, and other structures line the hillside. *Courtesy of Special Collections, Massachusetts Statehouse Library.*

This view of the west portal indicates the extent of brick arching necessary to hold back unstable "porridge stone" and control flooding. The women on top of the arching are probably tourists. The tunnel became a popular tourist destination during its construction. *Courtesy of Special Collections, Massachusetts Statehouse Library.*

Undated portrait of Walter Shanly. In late 1868, Walter Shanly and his brother Francis, both Canadian engineers, agreed to complete the Hoosac Tunnel for just under $5 million. Walter became the face of the partnership. He would benefit from the technology in place since the Thomas Doane era. *Courtesy of the McCord Museum, Montreal, Canada.*

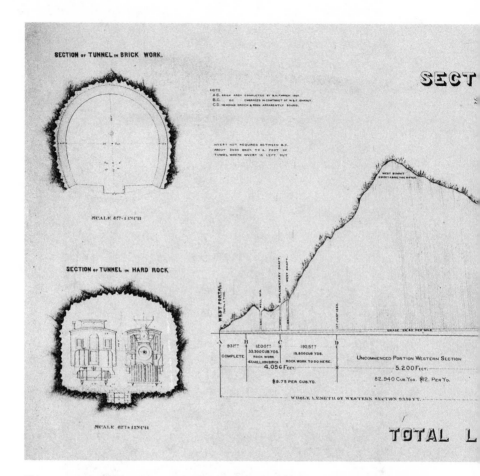

This cross-section of the Hoosac Tunnel shows progress by January 1869, just as the Shanly Brothers took over work there. Less than forty percent of the tunnel had been excavated. Had the Shanlys not agreed to complete the tunnel, it probably would have been moth-balled. The unfinished central shaft is visible in the middle of the cross-section. Smaller shafts sought more solid, stable ground at the tunnel's problematic western end. *Courtesy of Special Collections, Massachusetts Statehouse Library.*

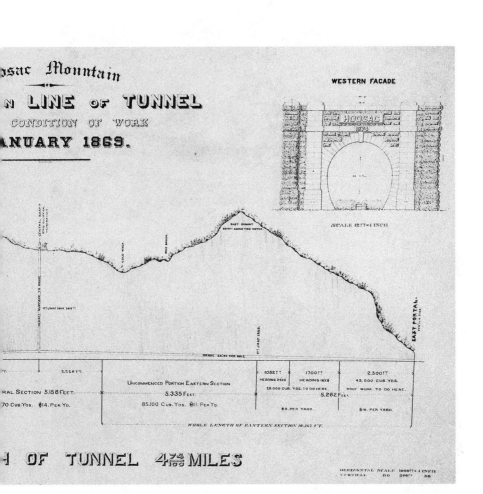

WESTERN FACADE

HOOSAC
1874

SCALE 12 FT = 1 INCH

CENTRAL SHAFT
FULL ELLIPSE

N COLD RIVER

RICE BROOK

EAST SUMMIT
HEIGHT ABOVE TIDE WATER

HOOSAC SURFACE TO GRADE

1ST JAN'Y 1869 560 FT

1ST JAN'Y 1869.

EAST PORTAL

GRADE 26.40 PER MILE

2.226 FT.   UNCOMMENCED PORTION EASTERN SECTION   1082 FT   1700 FT   2.500 FT
                                                                                HEADING 24X8 HEADING 16X8   45.000 CUB. YDS.

RAL SECTION 5.158 FEET.   5.335 FEET.   28.000 CUB. YDS. TO DO HERE.   ROOF WORK TO DO HERE.

70 CUB. YDS. $14. PER YD.   85.100 CUB. YDS. $11. PER YD.   5.282 FEET.

                                                                     $5. PER YARD. $16. PER YARD.

WHOLE LENGTH OF EASTERN SECTION 10.167 FT.

HORIZONTAL SCALE 1000 FT = 1 INCH
VERTICAL   DO   500 FT   DO

Entrance to George Mombray's tri-nitroglycerin factory, shown in undated photograph. Nearly half a million pounds of tri-nitroglycerin were produced by the English chemist George Mombray in his "acid house" near the west portal. *Courtesy of Special Collections, Massachusetts Statehouse Library.*

An expensive dam across the Deerfield River provided power for the air compressors that drove the Burleigh pneumatic drills. However, the river often ran too low to drive the air compressors. Both the dam and central shaft raised the cost of the Hoosac Tunnel and were unnecessary. *Courtesy of Special Collections, Massachusetts Statehouse Library.*

Brick kiln at west portal. One of Walter Shanly's biggest problems was cave-ins in the western portion of the tunnel. By the time of the tunnel's completion, over twenty million bricks would be produced at this kiln. More than seven thousand linear feet, or a quarter of the tunnel's length, would have to be arched. *Courtesy of Special Collections, Massachusetts Statehouse Library.*

The Rush from the New York Stock Exchange on September 18, 1873, by Howard Pyle, from *Scribner's*, July 1895. Just a month before the final breakthrough at the Hoosac Tunnel, the Panic of 1873 hit the country like a bombshell. It would usher in five years of economic depression and mute the economic promise of the tunnel. *Courtesy of the Delaware Art Museum, Wilmington, DE.*

The Railroad Committee of the Massachusetts legislature at west portal facade, 1875. Elaborate facades were installed at both ends of the tunnel, bringing the total length of the tunnel to 25,081 feet, or 4.75 miles. It was the longest tunnel in the Western Hemisphere. Still, the legislature's ill-advised "toll-gate" plan doomed the tunnel to insolvency until the Fitchburg Railroad purchased it in 1887. Only then did the Hoosac Tunnel fulfill some of the expectations for it. *Courtesy of Special Collections, Massachusetts Statehouse Library.*

4

HERMAN HAUPT AGAINST THE MOUNTAIN

HERMAN HAUPT SEEMED tailor-made to take on the Hoosac Mountain. He was born in 1817, and his generation rose to adulthood just in time for America's railroad boom. His father was a successful dry goods dealer in Philadelphia but died when Herman was eleven. The third oldest of eight siblings, he was thrown into a position of responsibility early and struggled to keep his mother and large family together. The experience may have instilled a degree of resentment in him. Supremely confident in his own abilities, he could be abrupt and tactless with those who disagreed with him. After graduating from the U.S. Military Academy at eighteen (he was thirty-first in a class of fifty-six), Haupt resigned his commission within a year to become a draftsman and then an engineer for a railroad running from Gettysburg, Pennsylvania, to Hagerstown, Maryland. By 1840, he had moved to the York & Wrightsville Railroad, where he conducted a number of experiments in bridge construction. During this time, he also taught mathematics at Pennsylvania College (later renamed Gettysburg College), where he earned a master's degree and wrote a book on railroad-bridge trusses. (It was published in 1851 as *The General Theory of Bridge Construction*.) In 1847, Haupt joined the Pennsylvania Railroad and worked on the Gallitzin Tunnel and Horseshoe Curve under Chief Engineer J. Edgar Thompson. Haupt rose quickly through the company's ranks, becoming chief engineer himself in 1853. Thompson became the railroad's president. Haupt gained further tunneling experience when called upon by the railroad to complete the 3,570-foot Summit Tunnel through the Allegheny Mountain. However, Haupt had been away in Mississippi for a year, having temporarily resigned during a board-room dispute, and came to the tunnel project when most of the actual boring had been completed. Nonetheless, Herman Haupt was now considered an expert in both bridge building and tunnel excavation.[1]

In photographs of him, Haupt is strikingly handsome but stiff and intense. His nose is strong and ruler straight. Dark eyebrows tilt upward from the bridge of his nose toward his temples in a flaring effect. Under them, his eyes are fixed and piercing. Haupt's features are perfectly balanced. Only a shock of black hair, combed diagonally across his forehead, violates their symmetry. He spoke with a booming voice and, though occasionally rancorous and impetuous, possessed a commanding presence. As a nondrinker with a religious bent, he could at times seem sanctimonious. While historians have correctly praised his role as "Lincoln's railroad man" in the Civil War—during which he repaired destroyed bridges with remarkable alacrity and kept Union railroads running smoothly—they have largely ignored Haupt's tenure as chief engineer on the Hoosac Tunnel in the years before the war.[2]

The story of how Herman Haupt came to the Hoosac Tunnel is complicated. Suffice it to say, Haupt had grown bored with his duties at the Pennsylvania Railroad once the line was up and running. He was more of a builder than a manager. As a consequence, he craved the kind of validation that would accrue to him as the engineer of the longest tunnel in America. There were also monetary considerations. At age thirty-eight, Haupt was moderately wealthy. He had invested in real estate, timber, and coal, property and commodities that he frequently sold on to his employer. He had also accumulated a significant amount of Pennsylvania Railroad stock and was elected to the company's board in 1855. Armed with his new prosperity and corporate standing, Haupt had joined a close-knit network of local capitalists, who often apprised each other of new investment opportunities. Ever restless and ambitious, Haupt was in search of ways to enhance his reputation and wealth.[3]

It is unclear whether Haupt was deceived by his business partners regarding the financial condition of the Troy & Greenfield Railroad or was naïve in his own assessment of it. Likely, both were true. As Edward Serrell, the railroad's chief engineer, edged nearer to bankruptcy and exhausted his sources of local financing, he reached out to a pair of capitalists named William Brown and William Galbraith from Erie, Pennsylvania, where he had worked in the 1840s. The latter became the key booster of the project, working closely with Serrell. In a letter to Samuel Lane, president of the North West Railroad that served

Erie, Galbraith duplicitously summarized the needs of the Troy & Greenfield: "We want men rather than money, management more than capital." The prize for investors, Galbraith explained to Lane, was an easy profit to be made from the $2 million in bonds approved for the Troy & Greenfield Railroad by the State of Massachusetts the year before, which could be discounted on the London bond market at 110 percent. In addition, there would be another $1 million profit when Serrell completed the Hoosac Tunnel for half the amount of the state aid package. Acknowledging his interest, Lane asked Haupt, who reputedly knew something about railroad tunnels, to visit the Berkshires and assess the opportunity firsthand.[4]

Haupt was also interested in the Hoosac Tunnel and traveled to North Adams, Massachusetts, to learn more. Incredibly, Haupt did not visit the actual tunnel but rather the so-called Little Hoosac Tunnel, an approximately four-hundred-foot bore through a large rock outcropping west of the town, blocking the railroad's intended path northward. The Stanton brothers, contractors from North Adams, were hard at work on the project and nearly done with it. The town was in an ebullient mood. The local newspaper was predicting the Hoosac Tunnel would be finished next. Impressed by the local enthusiasm, Haupt failed to ask the hard questions about the tunnel and the railroad that would go through it. Had he done so, he would have learned more. For example, after the Massachusetts legislature decided to allow towns along the Troy & Greenfield line to invest as much as 3 percent of their municipal valuations in the railroad—and after Serrell, Crocker, and Derby aggressively solicited these municipalities for stock subscriptions—the tunnelites failed to raise much of anything from these townships to fund the railroad's continuing construction. Once again, local enthusiasm for the Hoosac Tunnel had not translated into local stock purchases. At the same time, area newspapers lamented the unwillingness of Boston investors to help finance the new cross-state railroad, allegedly so vital to that city's commercial future. Haupt would also have learned about the tragic labor dispute at the east portal the previous month. Referred to in local headlines as the "Tunnel Affray," this pitched battle between rival work crews resulted in numerous wounded laborers and two crushed skulls. The incident underscored the desperation of poorly paid Irish immigrants at the site and the deplorable conditions under which they worked. The Stanton brothers were implicated in the incident and would cease working for the Troy

& Greenfield Railroad. Most important, Haupt did not investigate Edward Serrell's true financial condition, nor did he ascertain how deeply in debt Serrell and the Troy & Greenfield really were. In short, Haupt's visit to Massachusetts was a very cursory one. Upon his return to Pennsylvania, he informed Lane that "with the state loan, the payment of the subscription already made, and an additional subscription from Boston of only one-twentieth as much as Philadelphia had given in a case precisely similar, the whole work would be financially provided for and leave a large profit for the contractors." Such sloppiness on Haupt's part and dishonesty by Galbraith did not portend well for the new investors.[5]

Both Lane and Haupt immediately pledged $20,000 each to the Troy & Greenfield Railroad, and, in early January 1856, the firm of Serrell, Haupt & Company was formed. There were five partners—Serrell, Haupt, Galbraith, Brown, and Lane, each expected to invest equal amounts to provide the railroad with $100,000 of new capitalization. A contract was executed assigning responsibility for completion of the railroad and tunnel to the partners' new construction company for the sum of $3,570,000. It also stipulated that the railroad had to raise $210,000 in paid-in stock subscriptions by mid-May 1856, to help pay for work until the first state loan disbursement would be received. Within the month, however, this contract had to be abrogated because it was discovered that Serrell's debts exceeded the amount the Pennsylvanians intended to invest in the tunnel. Serrell was completely broke and in no position to contribute new money. Worse still, Haupt learned that his Pennsylvania partners—Brown, Lane, and Galbraith—could not make good on their commitments either. Lane's railroad was already in financial trouble and would go into receivership a few years later. Although he had paid nothing, Galbraith stayed on because he was serving as the company's attorney. When the dust settled, Herman Haupt alone remained fully invested in the Hoosac Tunnel.[6]

At the end of January 1856, a new contract was signed relieving Haupt of responsibility for Serrell's debts. Still, if Haupt was to carry on with the project, these debts had to be dealt with to preserve the railroad's credit rating. He also recognized that he could not go it alone without a fresh infusion of capital. During the months that followed, Haupt contacted a group of Pennsylvanians that he knew were more reliable and financially sound than those that had just abandoned him. First, Haupt arranged a $100,000 loan from two fellow board members of the Pennsylvania Railroad and a wealthy Philadelphia merchant named Horatio Burroughs. Haupt put up $321,000 of his own securities as collateral. After that, Haupt sold an 11 percent interest in his new construction contract with the Troy & Greenfield for $60,000 from a consortium of five investors that included his former boss at the Pennsylvania Railroad, J. Edgar Thompson, and a second participation by Horatio Burroughs. Finally, after forcing Edward Serrell out of his job as chief contractor of the Troy & Greenfield, Haupt brought in the Philadelphia engineering firm of Dungan, Cartwright & Company to replace Serrell. Although this firm specialized in urban gas and water works, it had $87,000 to invest in the Hoosac Tunnel. The partners in the firm were Charles Dungan, Henry Steever, and Henry Cartwright. Haupt lent the firm $125,000 in Troy & Greenfield bonds to serve as collateral for their loan. Remarkably, Haupt had managed to drum up almost a quarter of a million dollars in new financing from his Pennsylvania connections. He had saved the Troy & Greenfield Railroad from almost certain bankruptcy, at least for now.[7]

Herman Haupt was out of his element in Massachusetts. The Troy & Greenfield Railroad was a very different enterprise and faced very different challenges than the Pennsylvania Railroad, where Haupt had spent the last decade. The Pennsylvania Railroad was one of the best-managed companies in America. Its president during Haupt's time there, J. Edgar Thompson, as well as Thomas Scott after him, were rarities in this pre-professional era of business management. Furthermore, Haupt's former employer exercised a nearly monopolistic grip on the area it served and enjoyed robust support from its state legislature. The intense political combat between the tenaciously defensive Western Railroad and the fledgling Troy & Greenfield Railroad was a new field of play for Haupt. The more problems the Troy & Greenfield encountered trying to establish itself, the more ruthlessly the Western attacked it. The Western's president, Chester Chapin, and his chief political henchman, Daniel Harris, saw the new railroad's weakened condition in the mid-1850s as a chance to destroy their competitor once and for all. They were soon joined by Francis Bird, a wealthy papermaker and political wire-puller from Walpole, whose poison pen would produce a stream of pamphlets against the Hoosac Tunnel. These defenders of the Western monopoly made their attack personal by tarring Haupt as a corrupt and opportunistic "outsider," intent upon robbing the people of Massachusetts of the $2 million they had pledged to the Troy & Greenfield Railroad. Haupt would forever blame the opposition of the Western Railroad and not the Hoosac Mountain for his eventual failure in Massachusetts.[8]

Often overlooked by historians of the period, Francis William Bird played an important role in the Bay State's antislavery movement and was by far the most indefatigable opponent of the Hoosac Tunnel. Born in 1809, he was of Scotch-Irish descent and son of a prosperous paper manufacturer. He attended private academies followed by Brown College, where he studied moral philosophy before graduating in 1831. He was deeply religious and somewhat eccentric. He adopted the healthful regimen of Sylvester Graham, eating two meals a day, avoiding meat altogether, and drinking only cold water. He was tall but stooped and wore gold-trimmed spectacles for near-sightedness. He was interested in social reform and taught school for a time in one of Boston's most impoverished neighborhoods. A week after marrying Rebecca Cooke of Providence, he shocked his hometown Congregationalists by not only bringing his wife's black servant girl to Sunday service, but ushering her into the family pew before he and his wife entered it. Initially a Conscience Whig, Bird joined the Free-Soil Party in 1846. Later, Bird would become a staunch antislavery, or "radical," Republican. He would found the "Bird Club" and serve as the *eminence grise* behind several Massachusetts governors. Bird was a highly successful businessman and used the Panic of 1837 to buy up rival paper mills. In many ways, he and Alvah Crocker were different sides of the same coin.[9]

The political climate in Massachusetts had changed markedly since spring 1854, when the state's last Whig legislature approved the $2 million tunnel loan. That fall, a populist rebellion of unprecedented proportions swept through the state. A new party, officially called the American Party but soon referred to as the "Know Nothings," captured all but a handful of the state's 419 House seats, the entire Senate, and the governorship. The Bay State's political elite watched with disbelief as Henry J. Gardner, a Boston wool merchant and ex-Whig, occupied the governor's office. He would serve three terms. The Know Nothings revolt was a full-throated rejection of the political establishment and what it stood for. The *Springfield Republican* summarized the significance of what had happened: "Know Nothingism has done its work. It has crushed out the opposition to the establishment of a republican party. It has taught men who thought they had a prescriptive right to govern Massachusetts the depth of their error. It has cleared away the rubbish in the political field, and popularized politics." Not only could the tunnelites no longer count on Whig legislative support, but the new political order was a cipher.[10]

The Know Nothings organized in semi-secret "lodges" across the country and burst on to the national scene in the election of 1854. They elected eight state governors, over a hundred congressmen, and thousands of local officials. Boston, Philadelphia, and Chicago suddenly had Know Nothing mayors. The Know Nothings were anti-Catholic, antiforeign, and anti-immigrant but could also be antislavery, prohibitionist, and politically reformist depending on the local issues where they were elected. Most Know Nothings had little or no political experience. They were "new men," free of the taint of political parties and presumably more representative of the people. The Know Nothings scored stunning victories in New York, Connecticut, Pennsylvania, and elsewhere— but nowhere else did they win as resoundingly as in Massachusetts.[11]

Massachusetts was the perfect storm for Know Nothingism. Virulent anti-Catholicism had arrived early with the Great Migration. At the center of the Puritan project had been the struggle to rid society of the perceived impurities and depravity of the Church of Rome. During the eighteenth century, Bostonians celebrated Pope's Day every November fifth with large parades, attacks on Catholic neighborhoods, and the burning of papal effigies. Such sentiments demonstrated their durability with the burning of the Ursuline Convent in Charlestown, Massachusetts, in 1834. The massive immigration of Irish Catholics to Massachusetts after the 1845 potato famine only exacerbated these animosities. Early Bay State nativists blamed rising pauperism and crime on Irish immigration and accused Catholics of seeking to undermine republican freedoms in league with the papal officiate. Massachusetts Know Nothings blamed Irish voters for helping defeat temperance legislation and the Constitution of 1853. They reminded voters that Irish militia units had enabled the authorities to arrest fugitive slave Anthony Burns in spring 1854 and return him to his southern master. Because the Irish voted in large numbers for the Democratic Party and that party traditionally defended southern slavocracy, Know Nothing rhetoric conflated the threat of "slave power" with "Romish conspiracy." This proved a potent strategy in a state enmeshed in the slavery debate, rabidly anti-Catholic, and deeply suspicious of its rising immigrant population.[12]

Irish immigration altered more than just the politics of Massachusetts. By the late 1850s, it had destroyed Francis Cabot Lowell's dream of industrialization without the degradation of labor. The severity of the Irish potato famine and cheap fares from Liverpool to Boston—as low as $17 per person including provisions—inundated what had been a relatively homogeneous society with successive waves of unskilled, impoverished newcomers. Early arrivals found work on low-paying public and private construction projects where their brawn was in demand. Soon, however, immigrants entered factory work. Gradually, Francis Cabot Lowell's so-called Waltham model began losing traction, as Bay State factory owners substituted less expensive Irish males for native New England women. Before long, male factory operatives outnumbered women, and wages collapsed under the pressure of burgeoning immigration. This abundant and ever-increasing labor supply gave Massachusetts a competitive advantage versus New York. Whereas a New York factory worker was paid between $8.00 and $10.00 per week, a comparable worker in Massachusetts earned from $4.50 to $5.50. In desperation, some Irish workers and their families sought opportunities away from Boston and its environs. One of these was the Hoosac Tunnel. Nonetheless, the deadly "Tunnel Affray" in North Adams, Massachusetts (just before Herman Haupt's visit there), showed how scarce jobs were everywhere and that workers often had to fight to keep them. In the end, the steady supply of Irish labor arriving in Boston depressed wages at the tunnel too.[13]

In spring 1856, the tunnelites approached the new Know-Nothing-dominated legislature for the first time. Herman Haupt was still negotiating his Pennsylvania financing, and the tunnelites were desperate. They were requesting that the state purchase $150,000 in Troy & Greenfield stock, in exchange for several seats on the company's board. Massachusetts was a small state and "naturally sterile," Elias Hasket Derby began his argument before a joint committee of the legislature. "Poor in agriculture," Massachusetts "reaches out her arms to the prolific regions of the West." Due to its paucity of natural resources, the state had "devoted much attention to the cultivation of the mind." As a consequence, it had become preeminent in the art of manufacturing. Were it not for the failure of the Western Railroad, Massachusetts would be enjoying a robust exchange of its manufactured goods for western foodstuffs. The Western Railroad, Derby continued, was the result of impatience and not clearly thought out. Its steep gradients and tight curvatures, its "devious course" and "swinging around, up and down" exhausted the power of stream. Loammi Baldwin had identified the true path years before, a straight line from the Erie Canal through the Hoosac Mountain to the Port of Boston. Although he presented an extensive analysis of the relative efficiencies of the two railroads, Derby seemed intent on convincing the legislature that the state's mission was to straighten the "deviant" line of the Western Railroad: "If, gentlemen, we can, by the exertions of intelligence and science, break through this mountain, we are placing the channel of communications where it ought to be. We ought to be upon the great line of communication between the East and West." Derby was invoking the era's reigning faith in man's technical ability to reengineer nature, correct its irregularities, and make it better serve his needs. The straighter line of the Troy & Greenfield Railroad took on a symbolic, even moral dimension.[14]

It was not to be. The legislature voted down any state subscription of Troy & Greenfield stock. While newspapers in North Adams and nearby communities vilified the Western Railroad for again lobbying to defeat the tunnelites, such lobbying efforts were hardly necessary. As a populist movement, the Know Nothings were distrustful of state aid to private corporations, especially railroad companies. It smacked too much of what they saw as old Whiggery and its shameless collusion between government and business. The year before, Henry Gardner, the Know Nothing governor, had vetoed a similar aid bill to the Vermont & Massachusetts Railroad. He explained that "the state cannot make loans except in the case of great exigencies." Some states had enacted laws prohibiting state aid to railroads after the Panic of 1837. Others had followed suit by the 1850s. Massachusetts Know Nothings took the view that the state's railroads had ridden roughshod over the people. The new party advocated warnings at railroad crossings, restricted train speeds in urban areas, and other safety measures to protect the public from the railroads. If they were going to survive in this new political climate, the tunnelites needed a new strategy.[15]

That strategy was to petition the legislature to restructure the pay-out requirements of the $2 million loan, approved three years earlier but from which the Troy & Greenfield Railroad had not yet received a penny. The tunnelites were optimistic because the loan had been passed into law before the Know Nothing ascendency and the reasons for modifying it made sense. Haupt, Crocker, and Derby argued that the requirement for a thousand feet of tunnel and five miles of railroad track—necessary before each payment of $100,000—

should be decoupled and broken into smaller work and payment increments. In this way, the Troy & Greenfield could lay track that benefited the communities that needed it most and work on the tunnel separately. Smaller, more frequent payments would help the railroad's cash flow. Naturally, the Western Railroad attacked the restructuring petition aggressively. Daniel Harris spearheaded the attack, impugning Haupt's honesty, reading from falsified newspaper articles critical of the tunnel, and calling for examination of the Troy & Greenfield's bookkeeping. When all else failed, Harris encouraged legislators to visit the tunnel. This tactic backfired when these legislators returned with a positive report on Haupt's organization and progress on the tunnel. Both the Massachusetts House and Senate approved the restructuring bill by substantial margins. The tunnelite victory was short-lived, however, when Governor Gardner vetoed the bill in late May 1857. He characterized the tunnel as a "preposterous scheme" and one that would not be finished during the lifetime of anyone currently alive. Gardner cited his responsibility to the state and refused to "plunge her into incalculable expenditures with open eyes." An attempt to override Gardner's veto failed narrowly.[16]

While the Western's influence weighed heavily on the restructuring debate and attempted veto override, Gardner's veto itself was driven by shifting political circumstances. To understand them, some background is helpful. By 1857, Gardner and his Know Nothing Party were fighting for their political lives in the face of yet another new party—the Republicans. Although he had won reelection in 1855, Gardner's support had declined by a third versus his landslide victory the year before. Furthermore, the national Know Nothing Party had irreparably divided over the issue of slavery at their national convention and ceased to exist. (Only the southern "rump" of the Know Nothings would survive and support Millard Fillmore as the party's 1856 presidential candidate.) Regarding the Massachusetts party, it had lost face due to its foolish anti-Catholic measures and ambivalent stance on slavery. Too often state Know Nothings promoted their bizarre nativist agenda at the expense of antislavery measures deemed more important by voters. For example, local nativists removed Latin inscriptions from the State House walls and set up a "nunnery commission" to investigate alleged depravity in convents. For his part, Governor Gardner gave only lukewarm support to abolitionist efforts to protect escaped slaves in the state and refused to aid antislavery forces fighting in

"Bloody Kansas." Nonetheless, Gardner's political machine was still stronger in Massachusetts than that of the Republicans. In the presidential election of 1856, Massachusetts Republicans offered to back Gardner for governor in exchange for Know Nothing support of John Fremont, the Republican's first-ever presidential candidate. This unusual deal was negotiated by Massachusetts congressman Nathaniel P. Banks, former Democrat turned Know Nothing and now morphing into a Republican. While Fremont won the Bay State, he lost the presidency to Democrat James Buchanan. Gardner secured a third term but anticipated that Banks would challenge him in the 1857 gubernatorial race. By vetoing the loan restructuring bill for the Troy & Greenfield Railroad, Gardner curried favor with conservative Bostonians who favored "retrenchment" in state spending and with those towns in the southern part of the state loyal to the Western Railroad. Many Know Nothings also applauded his fiscal restraint. Always the student of shrewd political calculus, Gardner believed he could sacrifice the less important towns in the northern half of the state. These towns were enraged at Gardner's veto and looking forward to the prospect of Banks running against him for governor. Known as the "Bobbin Boy of Waltham," Banks's humble background as a textile worker and vibrant style on the hustings made him extremely popular. Given that he hailed from Waltham, on the Fitchburg Railroad line, Banks would probably support the Hoosac Tunnel if he gained office. In short, the tunnel had become what modern pundits would refer to as a "political football." The tunnelites needed to wait.[17]

Still, 1857 turned out to be a difficult year. An economic contraction began in August, when it became clear that the end of the Crimean War would reduce European demand for American wheat. The downturn was most severe in the West, where land speculation and railroad building had gotten ahead of themselves. But its effects were soon felt in the East. Struggling to maintain liquidity, New York and other money-center banks were forced to tighten credit. Several of Haupt's notes were called and some of his Pennsylvania property seized by the sheriff. Haupt scrambled to assign his house and other property to friends and associates. He probably owed upward of $200,000 at the time and was lucky to survive the crisis. As expected, Nathaniel Banks won the fall gubernatorial race in Massachusetts by a substantial margin. Though Henry Gardner was gone, the Know Nothings were still a force in Massachusetts politics and were especially skittish about public spending following the financial crisis. Furthermore, Francis Bird and the antislavery extremists in the Republican Party hated Banks for his earlier collusion with the nativists and looked for any excuse to bring him down. Bird was especially vigilant regarding Banks's handling of the Hoosac Tunnel. Since no immediate volte-face could be expected on tunnel funding, Haupt turned again to his Philadelphia friends. He procured an unsecured, low-interest loan of $33,000 from Alexander Derbyshire, a former Pennsylvania Railroad board member. It was enough to continue work on the tunnel.[18]

Whereas Gardner's veto and the panic of 1857 had discouraged many of Haupt's Pennsylvania investors, Banks's victory in the Massachusetts governor's race and the brevity of the financial collapse helped restore confidence in Haupt's tunnel scheme in his home state. These sentiments likely accounted for Alexander Derbyshire's liberality. Also, Haupt was an excellent salesman. In early 1858, he unleashed his selling skills on the towns of northern Massachusetts. Per earlier legislation, these towns were authorized to invest up to 3 percent of their municipal valuations in Troy & Greenfield stock subscriptions. Unlike the lackluster fundraising drive by Crocker and Derby several years before, Haupt's effort was better organized, capitalized on his reputation as a can-do railroad man, and was accompanied by parades and pyrotechnics redolent of a political campaign. He led off with a newspaper blitz in the first months of 1858, describing his progress laying tracks and boring the tunnel, followed by town meetings in which Haupt spoke and the local citizenry debated and voted on the stock subscriptions. Town meetings were held in Williamstown, in the northwestern-most corner of the state; North Adams, next to the Hoosac Tunnel's west portal; Florida, a village sitting on top of the Hoosac Mountain; as well as Charlemont and Buckland, towns facing each other across the Deerfield River on the mountain's eastern watershed. Still smarting from Gardner's affront to their communities, these townships opened their pocketbooks in gratitude for Haupt's perseverance and to demonstrate their enthusiasm for the tunnel.[19]

The town meeting in North Adams on February 24, 1858, was typical. Haupt described his work on the tunnel to date and what challenges were ahead in clear layman's terms. He committed to the tunnel's completion if adequate funding were forthcoming. "Nearly a thousand voters were present and many who came with the intention of giving the road the go-by were converted to its favor by Haupt's remarks," the *Hoosac Valley News* reported. In the end, North Adams pledged $60,000 to support Haupt's work. After the vote was taken, the town celebrated with a 125-gun salute and bell ringing well into the night. In Williamstown, Mark Hopkins, the president of Williams College, joined Haupt at the podium and claimed that "all the college" was behind the tunnel. All in all, the towns where Haupt campaigned pledged $145,000. These endorsements were a welcome palliative after his discouragements before the Massachusetts legislature and close brush with financial disaster. Haupt's meetings also helped solidify political support for the Hoosac Tunnel in the towns he visited.[20]

In August 1858, the prize that had originally lured Haupt and his Pennsylvania investors to the Hoosac Tunnel—the $2 million Loan Act of 1854—was finally within reach. More precisely, the construction requirements for the first payment of $100,000—one thousand feet of tunnel and seven miles of track—had been met. Haupt alone remained to claim the prize. Edward Serrell was out of the picture, as were Haupt's original Pennsylvania partners. So, too, the Philadelphia construction firm of Dungan, Cartwright & Company had abandoned the tunnel during the financial panic of 1857. By now, Haupt qualified as the Troy & Greenfield's largest stockholder, its general agent, chief engineer, and treasurer. The fact that he occupied all these positions allowed him to circumvent the one loan requirement the railroad had not satisfied, the 6 percent paid-in stock subscription (the company had only $20,000 of paid-in subscriptions versus the $120,000 required). In an act of financial legerdemain, Haupt had his general contractor borrow $100,000 from a bank and transfer it to the Troy & Greenfield treasurer (Haupt, of course), who recorded it as paid-in stock subscriptions. Haupt then paid the contractor this same amount for services to the company, and the contractor returned the money to the bank. The entire transaction took one day. Surprisingly, the attorney general ruled the transaction legitimate. Though his enemies would later accuse him of defrauding the state, Haupt's questionable financial maneuver succeeded. Still, nothing having to do with the state loan was ever simple. Payment of the first installment was further delayed by several creditors who entered attachments against it in the courts. One was Elias Hasket Derby, to whom the Troy & Greenfield owed $3,000 in legal fees. Sadly, Haupt's terse, arms-length treatment of Derby cost the railroad one of its staunchest supporters.

By now, however, Haupt *was* the Troy & Greenfield Railroad. Its future depended on him.[21]

Fortunately for Haupt, receipt of the first loan payment in October 1858 marked the beginning of a political "sweet spot" for him. Nathaniel Banks would be reelected governor that fall and again the next year. Banks shrewdly balanced the demands of the Know Nothings with those of the newly emergent Republicans. He was unwavering in his advocacy of a state constitutional amendment barring immigrants from voting for two years after their naturalization. With the governor's support, the legislature and a state-wide referendum passed the two-year amendment into law. At the same time, Banks outflanked the radical antislavery Republicans by firing Judge Edward Loring, who had approved the rendition of fugitive slave Anthony Burns to his southern master several years earlier. Banks also supported so-called personal liberty laws to protect the rights of fugitive slaves in the state, as well as state aid to proslavery forces fighting in Kansas. Knowing that the governor was in favor of the Hoosac Tunnel, Haupt approached the State House again with a petition to modify the state loan requirements.[22]

Haupt's campaign began in earnest during January 1859. As usual, Daniel Harris played pointman in the Western Railroad's defense. Because so many legislators and journalists had visited the tunnel and lauded Haupt's apparent progress there, Harris changed his oppositional strategy. Now that the tunnel seemed feasible, he no longer talked about how long it would take to finish the tunnel but rather how Haupt was embezzling the state with his accelerated work schedule. Haupt was "going to get through the mountain so fast that he will be immensely rich," Harris argued. This shift in strategy was confusing for the public, given the Western's earlier argument that the tunnel would take decades to complete. Harris's accusations also lost credibility when Haupt debated him in front of the joint committee on railroads. Haupt presented his facts clearly and calmly, much as he had at his recent town meetings. When it came to a vote, the Senate came down 21 to 10 in favor of the loan modification bill. The House followed suit with a vote of 167 to 54. Governor Banks signed the bill into law at the end of March. The new bill set aside $700,000 for work on the railroad, apart from tunnel construction, and classified all town stock subscriptions as paid-in. Haupt would receive $25,000 for each five miles of completed railroad. Payments for tunnel work would be $50,000 for each thousand feet of finished bore. A sense of progress on the tunnel, support from adjacent communities, Haupt's own salesmanship, and a tunnel-friendly governor all came together to give Haupt his first legislative victory.[23]

Though a significant political victory, the loan modification bill brought no immediate money with it. By midsummer 1859, Haupt was again out of funds and forced to seek a $5,000 loan in Boston to meet his payroll. Then his financial circumstances began to change. Sympathetic to the high cost of driving the railroad up the Deerfield Valley, the Town of Greenfield subscribed to $30,000 of Troy & Greenfield stock in July. By September, a second thousand feet of tunnel had been penetrated and triggered a payment of $50,000 from the state loan. At the end of the year, another $25,000 was released for five miles of completed trackage east of the tunnel. By this time, Haupt had approached the legislature for a further modification of the state loan. Over intense opposition by the Western Railroad, the Loan Act of 1860 became law in February of that year. Among its many provisions, the new law would pay out the state loan in small, monthly increments for completed work on the tunnel and railroad. It also eliminated the paid-in-stock requirement. In exchange for dropping this requirement, a state engineer appointed annually by the governor would verify all finished work before each payment. Most fortuitous of all, the bill contained an offer by the State of Massachusetts to purchase the Southern Vermont Railroad, a fully owned and integral part of the Troy & Greenfield Railroad, for $200,000. This six-mile stretch of railroad ran across the southwesternmost corner of Vermont on the way from Williamstown, Massachusetts, to Troy, New York. The Massachusetts legislature was apparently uncomfortable with its future jurisdiction over this portion of what might one day become a functioning cross-state railroad. Since he was the general agent for the Southern Vermont Railroad, Haupt wasted no time selling it to the state and putting this financial windfall to work on the tunnel. Suddenly, it was raining money.[24]

Still, no amount of money was enough to breach the Hoosac Mountain. Progress was too slow and need for money seemingly never-ending. Though later engineers would make important advances, the tunneling technology Haupt desperately needed was not available to him in 1860. Despite the public's prevailing sense of progress, only 2,300 feet had been bored by August of that year through a tunnel that would eventually measure more than 25,000 feet (or 4.75 miles) in length. After almost a decade of struggle, less than a tenth of the Hoosac Mountain had been penetrated. What is more, hardly any progress had been made at the western end of the tunnel. Only a few hundred feet had been driven through unmanageable "porridge stone," or "demoralized rock," as the tunnel's critics liked to call it, at an unsustainable cost of life and money. While the main body of the Hoosac Mountain had been formed in ancient times, its western side was a kind of "foothill," some three-quarters of a mile wide and 250 feet deep, composed of more recently accreted clay, quicksand, and rock that disintegrated when exposed to air and light. Acknowledging the problem, Haupt dug a vertical shaft a few hundred feet east of the tunnel's west portal to establish a new heading in more solid ground. After failing to find stable material there, he sank a second shaft further up the mountainside. Haupt finally located solid rock there, at a depth of 318 feet. Workers at the bottom of the shaft began digging in both directions. A pair of mules were lowered into the darkness to haul excavated material and a bucket-hoist system was installed to bring tailings to the surface. A small camp for the workers grew up around the top of what became known as the "west shaft." Haupt also installed extra timber bracing at the west portal to help contain the porridge stone and reduce the number of cave-ins. All of this was hugely expensive and had yielded less than five hundred feet of progress at the western end of the tunnel by 1860.[25]

Isaac Browne was a boy living on his father's farm next to the west portal and became a spectator to these events. Years later, he recalled what he had seen. Most of the tunnel workers were from Ireland and spoke only Gaelic. This made them doubly strange to the young boy. The workers and their families lived in primitive shanties, built into the hillside in three rows. The outer walls of the shanties were banked with sod up to their eaves for warmth, leaving only a door for access and a single window for light. Since funding for the tunnel was intermittent, the tunnel workers frequently went without pay and had little to eat. Some stole apples and berries from the Browne farm and nearby properties. Most locals showed their sympathy by looking the other way. They were aware of the terrible accidents at the west portal. During early 1857, an explosion fractured the skull of Daniel Fourney, a thirty-six-year-old native of County Cork, and, a few months later, falling rock crushed Patrick McArty, his twenty-three-year-old countryman. The following year, two more Irishmen, John Shean, age thirty, and Murty McCarty, of unrecorded age, died from falling rocks when a loose gravel seam collapsed. Some miners did not die right away and were ministered to by Isaac Browne's uncle, Dr. Babbit. Since there was no hospital in North Adams nor anywhere else near the tunnel, Babbit did the best he could by candlelight in the miners' shanties. Given the state of medicine at the time and severity of the injuries, there was little Babbit could do.[26]

Most progress was at the tunnel's eastern end, where the mountain consisted of more compact rock. Nonetheless, work there was also slow, at best no more than four feet per day. While much of the material there was readily penetrable mica schist, seemingly adamantine veins of gneiss and quartz could stop progress completely. Workers were limited by antiquated tunneling methods that had not changed in a century. Two or three-man crews attacked the rock heading inside the east portal. Often, two crews worked side by side. They manually pounded what were known as "star drills" into the rock face before them. The "strikers" wielded twelve- to fifteen-pound hammers to force the drills into the rock, while "holders" rotated the drills after each hammer blow. (In some accounts, these men are referred to as "drivers" and "shakers.") Once a two- to three-foot-deep hole had been drilled, it was stuffed with black powder and a fuse was lit. After the blast, shattered rock was loaded into carts and hauled out of the tunnel by mules. It was dangerous work in close quarters. One errant hammer blow could break an arm. Worse still, black powder and the fuses available at the time were notoriously unpredictable. Blasting accidents were frequent and often fatal. In December 1859, a gang foreman at the east portal, John Fennel, checked a loaded explosive charge that had "hung fire," only to have the upper part of his head blown off. He died twenty-four hours later. Early the next year, Edward Rhule was badly mutilated in a similar accident. Miraculously, he survived. Another hung-fire explosion soon after blinded a miner remembered simply as Cameron. The deeper the tunnel progressed, the more stifling the atmosphere became. Herman Haupt installed a crude ventilation system using a cloth tube hung from the tunnel ceiling. A mule-powered fan at the entrance forced air into the cloth tube and tunnel. While this made work in the tunnel more comfortable, it did nothing to mitigate its danger.[27]

Haupt's ingenuity did not extend to mechanical drilling equipment, the most pressing need at the Hoosac Tunnel. His first effort was a large boring machine designed by Charles Wilson, the same inventor who had earlier fathered the Excavator, the rusting carcass of which still sat next to the east portal. While somewhat smaller at forty tons, Wilson's second drilling machine consisted of a massive head of rotating blades intended to cut an eight-foot-diameter hole in the tunnel's rock heading. After nine months of adjustments and modifications, this contraption was tested in late August 1857. Predictably, it met the same fate as its big brother and had to be scrapped. The failure cost Haupt $25,000 he could ill afford. Another costly failure involved his employment of South Boston mechanic Stuart Gwynn. Haupt thought Gwynn quirky but collaborated with him for two years trying to perfect a smaller steam-powered percussion drill, to replace the muscle-power of the strikers and holders in the tunnel. Their drill design was based on a set of patents registered by Philadelphians Jonathon Couch and Joseph Fowle, which, though promising, had never been tested in an actual tunneling operation. When Gwynn's foundry burned to the ground, destroying his only prototype, this project too ended in failure. Again, Haupt was forced to write off his investment of time and money. In addition to specific design flaws, these experiments traced their failure to the poor quality of iron and steel available at the time. None of these drills could stand up to the hardness of the rock in most tunnels. Nonetheless, Gwynn's design and those of the Philadelphians before him were on the right track. Behemoths like Wilson's drilling machine and other mammoth contraptions would give way to smaller, carriage-mounted, steam-powered or compressed-air drills. Gwynn would eventually rebuild his prototype but too late to help Haupt penetrate the Hoosac Mountain. Still, Haupt became obsessed with steam drills. He continued to work with Gwynn on his design, advocate for its deployment at the tunnel, and engage in lengthy patent fights to protect it.[28]

One of the backstories of this period was the reappearance during 1860 of several of Haupt's former partners, demanding a share of the state loan payments as well as dividends from the Troy & Greenfield Railroad. Although they had abandoned Haupt by the financial panic of 1857, they returned armed with their original contracts, in some cases shares of the company's stock, and the knowledge that Haupt had fresh money from the state. The most dogged were William Galbraith, one of the original five investors in the tunnel project, and Charles Dungan and Henry Steever, two partners in the Philadelphia construction firm that had replaced Edward Serrell. After multiple meetings with Haupt, these men were able to extract tens of thousands of dollars that Haupt badly needed to cover his payroll and later regretted paying them. Since his bookkeeping was a shambles, Haupt was unable to show where the railroad's money had gone and was open to accusations of embezzlement. Wretched accounting was one of Haupt's signal failures as manager. Though he had worked at the Pennsylvania Railroad for J. Edgar Thompson, whose grasp of accounting was legendary, Haupt had apparently learned little from him.[29]

But why, if progress on the Hoosac Tunnel was so lagging and its finances so precarious, did people not recognize this? The answer had to do with both Haupt and the nature of the tunnel. Regarding the tunnel's finances, there were no controls over how Haupt spent money. He handled almost every aspect of the Troy & Greenfield Railroad himself. And while Haupt was largely dependent on state money after 1858, the state had little control over what he did with it. The state only belatedly assigned an engineer to verify completed work for the loan payments. Regarding progress on the tunnel, people knew only what they heard and saw. By 1860, even representatives of the Western Railroad were convinced that work on the tunnel was speeding toward completion. After all, Haupt had said so. His reputation as an engineer, commanding presence, and cool presentation of what he said were the facts made him utterly credible. This credibility was by far his strongest suit, though he used it to perpetrate a colossal falsehood. At a rail-laying ceremony in North Adams during the Fourth of July 1858, Haupt looked eastward and said he "could see daylight through the Hoosac Mountain." Whether he believed such a prophecy was about to be fulfilled is unimportant. His audience believed it. Once the rail link to Troy, New York, was completed the next year, the textile mills of North Adams began to flourish due to lower freight rates and accessibility to new markets. Railroad tracks were something people could see and from which their communities benefited. It was easy to impute from such evidence the imminent completion of the entire railroad, including the tunnel. The *Hoosac Valley News* declared that "few intelligent persons see what is already done without concluding that the whole is possible." Of course, the tunnel itself was more difficult to fathom, even for those with access to it. When he visited it in the summer of 1860, Governor Nathaniel Banks was lowered into the tunnel's still unfinished west shaft and then toured the tunnel's longest bore from the east portal. He was treated to a black-powder detonation in his honor. He emerged, shaken, to wild huzzas from the workers but he was no wiser regarding the tunnel's actual progress. For Banks and almost everyone else, the tunnel was an enigma—complex, subterranean, and inscrutable.[30]

With most of his financing now coming from the state, Haupt was more vulnerable than ever to the vicissitudes of Massachusetts politics. While Nathaniel Banks had won over the state's nativists with his support of the so-called two-year amendment delaying the voting rights of newly arrived immigrants, it ultimately proved a Pyrrhic victory. The radical, antislavery wing of the Republican Party mocked Banks for his nativist kowtowing and saw the amendment as a blemish on the party. The radical Republicans also criticized Banks for his reluctance to enroll blacks in the militia, his willingness to allow Daniel Webster's statue on State House property, and the administration's censorious view of the Harper's Ferry incident. As the nation moved closer to Civil War, the antislavery radicals easily dominated the Massachusetts Republican Party and gave the 1860 gubernatorial nomination to John A. Andrew. The chief power broker behind Andrew was Francis Bird, leader of the radical "Bird Club" and friend of the Western Railroad. Once again, the political ground in Massachusetts had shifted.[31]

Initially, Haupt was worried about the requirement in the new loan act that a governor-appointed state engineer needed to approve his work before each loan payment. But he found the state engineer, Colonel Ezra Lincoln, highly professional and easy to deal with. Lincoln had accompanied Governor Banks on his August 1860 visit to the Hoosac Tunnel and been similarly awestruck by the experience. This happy relationship was fleeting, however. Circumstances changed when John Andrew won the fall gubernatorial election and Lincoln became too ill to carry on as state engineer. Sensing trouble ahead, Haupt wrote a letter of resignation for Lincoln to sign, which was to include a recommendation that his assistant Charles Stevenson replace him, and Haupt delivered the letter personally to outgoing Governor Banks for approval. Haupt even attended the governor's council meeting to ensure Stevenson was securely installed. Haupt also managed to get the November loan payment of $130,000 through Banks's office and he submitted a request for the December payment. After Andrew's inauguration in January 1861, the new governor expressed his resentment at Haupt's eleventh-hour machinations by refusing to approve the

December loan payment, which Banks had neglected to sign before leaving office, and demanding Stevenson resign as state engineer. Although the governor's council gave Stevenson until June to serve out the balance of Lincoln's term and recommended state loan payments continue based on the state engineer's verification, serious damage had been done to Haupt's relationship with Governor Andrew. Francis Bird was quick to exploit the rupture between the two men. Bird had the ear of the governor who he had helped place in office and was the sworn enemy of the Hoosac Tunnel.[32]

During early 1861, in addition to ongoing work at the tunnel, Haupt was attempting to complete the thirty-mile stretch of railroad between the east portal and the town of Greenfield. Besides the tunnel itself, this railroad up the Deerfield Valley was the most challenging construction he had faced. Purchasing the right of way was expensive and the terrain up the narrow, winding valley difficult to traverse. Unfortunately, Haupt was badly distracted by his continuing negotiations with his ex-partners and political struggle with the Andrew administration. During the first half of 1861, he continued to receive state loan payments. Still, these payments were insufficient to cover his construction costs and another dividend to his ex-partners. Haupt had been an infrequent visitor to the work site in the Deerfield Valley when a bridge he had designed over the Green River collapsed in mid-April, killing one workman and badly injuring two others. While the cause of the collapse was never established, the event was pivotal for Haupt's reputation. After all, bridge building was Haupt's stock-in-trade and the core of his engineering expertise. Unlike what was going on in the darkness of the Hoosac Mountain, the collapsed bridge over the Green River was a disaster in broad daylight for all to see. Bird and other opponents of the Troy & Greenfield Railroad wasted no time exploiting the event. In June, Governor Andrew replaced Stevenson as state engineer with William Whitwell, a friend of the Western Railroad's Daniel Harris. Whitwell began drastically reducing state loan payments, based on what he claimed was poor-quality construction. Haupt began to starve financially. Unable to meet his payroll, he ceased work entirely on the tunnel during mid-July 1861.[33]

Naturally, the local citizens in North Adams and Greenfield were not privy to the details of the work stoppage nor knew how long it would last. Furthermore, it was overshadowed by the outbreak of the Civil War. Fort Sumter had come under fire from Confederate batteries in Charleston, South Carolina, during April 1861 and Governor Andrew had dispatched Massachusetts soldiers to defend the nation's capital. The Battle of Bull Run and its humiliating defeat of Union forces occurred within days of the work stoppage on the tunnel. Nonetheless, the people in the communities with an interest in the tunnel were not fooled. At the end of July, the *Hoosac Valley News* published an article entitled "The Enemy Unmasked." The newspaper quelled rumors that state had ceased work on the tunnel because of "the large demand on its finances for the war." Rather, the reason for the stoppage was a conspiracy in the State House to "destroy an enterprise which will rob the Western railroad of its monopolistic power." The article blamed state engineer Whitwell, who had "suffered himself to be catspawed by the sworn enemies of the enterprise." It accused him of outright collusion with Daniel Harris of the Western Railroad. By late October, the same newspaper was saying that Governor Andrew was also to blame. He had only saved himself "from the State sepulcher beside the bones of Gov. Gardiner by going in for the war." Still, by sustaining Whitwell's "tom-fool calculations in regard to the great work," the paper declared, Andrew had written off his political support in northern Massachusetts. As expected, the region punished Andrew severely on Election Day 1861. Chastened, he began to soften his opposition to the tunnel.[34]

With boring of the Hoosac Mountain and construction of the railroad up the Deerfield valley abandoned, Haupt and Whitwell began fighting it out in the Massachusetts legislature and the court of public opinion. During January 1862, the two appeared before a special legislative committee to argue their cases. While Whitwell insisted that Haupt's work had been flimsy and below accepted engineering standards, Haupt explained that he had built expeditiously, in order to place the railroad in operation quickly and thereby generate revenues for subsequent improvements. In fact, Haupt had done just that with the Summit Tunnel while working at the Pennsylvania Railroad. He had allowed trains to pass through the tunnel before it had been properly arched, using the proceeds from that traffic to fund costly arching later. As expert witnesses, Benjamin H. Latrobe, Jr., chief engineer for the Baltimore & Ohio Railroad, and W. H. Wilson, who held the same position at the Pennsylvania Railroad, vouched for the quality of Haupt's work and judged it in compliance with the standards of their own roads. Whitwell countered with his own witnesses. One prominent member of the anti-tunnel faction, Senator Charles Loring, defended Whitwell's assessment of Haupt's work as "shoddy" and pointed to the collapse of the Green River Bridge as irrefutable evidence of his incompetence as an engineer. Alvah Crocker leaped to Haupt's defense, reviewing Haupt's resume before coming to the tunnel and reminding the legislators that President Lincoln had just appointed him to the board of West Point. Haupt was fighting to salvage both his reputation and financial future. He maintained that he had spent $1,226,755 to dig 4,250 feet of tunnel and lay the railroad tracks to it. He claimed he was owed $361,521 for his personal expenditures. After two months of testimony and cross-examination, the special committee sided with Haupt and admonished Whitwell for forcing construction on the tunnel to shut down. The committee recommended that work on the tunnel recommence immediately, Haupt be retained as chief engineer, and he be reimbursed $150,000 by the state. However, before Haupt could celebrate, Francis Bird ruined his victory in a single blow.[35]

Bird struck with his pen. Among his several anti-tunnel pamphlets published during the 1860s, *The Road to Ruin, or, The Decline and Fall of the Hoosac Tunnel* was by far Bird's most effective. Published within days of the joint committee's exoneration of Haupt, it achieved immediate and widespread distribution. Its content was sensational and changed the political discourse on the tunnel and the public's perception of its former chief engineer. Bird began by declaring the entire tunnel scheme a magnificent hoax: "rose-colored sketches of teeming wealth of the West, pouring into Boston through an auger-hole fourteen feet wide and eighteen feet high, through the Hoosac Mountain!" While initially a private railroad with intentions to raise capital from the towns along its proposed route, the Troy & Greenfield Railroad had been innocent enough. Even after the railroad failed to raise that capital and turned to the state for a loan, there were significant safeguards protecting the public from possible abuse by the railroad's management. However, all that changed after Herman Haupt, an "irresponsible stranger to Massachusetts," took over the company. "The Troy & Greenfield Railroad is a myth," Bird declared. Haupt owned most of its stock and held most positions of authority. Furthermore, Haupt had convinced the state to give away all its financial safeguards and gained "absolute control of two million dollars of State script." All that was left in the Loan Act of 1860 was the role state engineer and the requirement that all work on the tunnel be "substantially performed." Now, Haupt was contesting the judgment of the state engineer and claiming that construction on the tunnel met that standard.[36]

Point by point, Bird demolished Haupt's contention that work on the tunnel and adjoining railroad tracks had been substantially performed. The steep grading of the roadbed would never stand up to New England winters, Bird argued. Haupt had sloped them at a one-to-one ratio (45 degrees) versus the one-to-one-and-a-half ratio (33 degrees) required. Nor would cheap trestle-work over ravines endure long, where earthen embankments were called for. Haupt had also failed to use rip-rapping (loose stone) where grading came in contact with rivers and streams. The collapse of the Green River Bridge had been caused by faulty design, not material failure as Haupt claimed. All in all, repairing his mistakes would cost Massachusetts taxpayers $113,000. Finally, Haupt had failed to erect fences, install switches, and meet other requirements to make the railroad operational. Taken together, Haupt had charged the state $282,659 more than construction to date was worth. On top of this, the legislature wanted to pay him what Bird called a "gratuity" of $150,000.[37]

In closing, Bird turned his invective on Haupt's character. He had come to Massachusetts looking for "El Dorado" and "leagued himself with the most unscrupulous political gamblers in the legislature and profligate adventurers in the state." Bird was at a loss to explain the man's ability to influence others. "His versatility is wonderful," Bird marveled. "Plausible and insinuating toward those whom he desires to use, servile and sycophantic toward those who have the power to forward or thwart his schemes, arrogant and insolent toward those he can neither humbug or buy." Bird urged his fellow citizens to see through Haupt's sanctimonious speeches and feigned disinterestedness in his own ambitions. Bird likened Haupt to the loathsome villain Fagin in Charles Dickens's *Oliver Twist*, a loaded comparison few could miss. (In the novel, Dickens repeatedly describes Fagin as a scheming Jew.) "Haupt's game has been to get all the money he could out of the state for the road and leave the tunnel in our hands," Bird declared. If the tunnel was to be finished, the state had to do it. "We know too little about it," Bird confessed. The state should thoroughly investigate the tunnel before restarting work on it.[38]

The publication of Bird's *Road to Ruin* clarified the tunnel question for Governor Andrew. He disliked Haupt personally but was ambivalent about the tunnel. He was quoted as saying "I regret encountering any questions on the subject." Still, Andrew worried about his political support in the towns directly affected by the tunnel and sought to protect his legacy should the tunnel eventually be built and become successful. After Bird's pamphlet, however, Andrew could not support the legislature's recommendation to retain Haupt as chief engineer, much less pay him $150,000 of state money. Additionally, getting rid of Haupt would curry favor with those towns in the southern half of the state that were friendly toward the Western Railroad. In the end, Andrew threatened to veto the tunnel bill and named a special three-man commission to investigate the tunnel. It was what Bird wanted and the only option for Andrew. Recognizing they lacked enough votes to override the governor's veto, the tunnelites settled down to await the commission's report. If the Hoosac Tunnel was to be completed—and it was not clear in the spring of 1862 that it would be— Herman Haupt would not be the man to do it. Francis Bird had seen to that.[39]

After seven years of struggle, Haupt agreed to surrender his tunnel contract to the state. According to the terms of the settlement, the state assumed most of Haupt's liabilities to his subcontractors and granted him the right to appeal to the state for any money he believed he was owed up to ten years after completion of the tunnel. Though these assurances were purposely vague, they were the best he could get. More important, Haupt was off to war. He would soon establish a brilliant reputation managing railroad logistics for the Union cause. Within weeks of being commissioned a colonel, Haupt was using the same expedient approach to repair destroyed railroad bridges and torn-up track that had brought odium down on him in Massachusetts. When President Abraham Lincoln rode across Haupt's newly rebuilt Potomac Creek Bridge, he marveled at the raw beauty of Haupt's improvised construction. Haupt had repaired the four-hundred-foot span in less than ten days with untrained soldiers using nothing more than, in Lincoln's words, "beanpoles and cornstalks." By August 1862, Haupt was in charge of all railroads serving the Army of Virginia in the primary theater of the war. A year later, Haupt played a critical role in transporting troops and supplies to Gettysburg. By that time, Haupt's remarkable bias for action and feats of logistical magic had earned him the rank of general.[40]

Nonetheless, the Hoosac Mountain threw a shadow over Haupt's military career. He never signed his military commission, refused to wear a uniform in the field, and took no salary serving his country. In this way, he believed he remained free to return to Massachusetts at any time to argue his interests in tunnel-related matters. These visits became more frequent after Governor Andrew's three-man commission rendered its report on the tunnel. The commission was not kind to Haupt, accusing him of cheating the state out of $324,872 and affirmed Bird's damning critique of his work on the tunnel. Although the commission recommended the tunnel be completed—now at an estimated cost of $5,719,330—Haupt was furious about his treatment by the commission. After several more visits by Haupt to the state, Governor Andrew had had enough. He arranged for Edwin Stanton, the secretary of war and Haupt's boss, to force his subordinate to officially accept his commission, thereby keeping him away from Massachusetts as long as the war lasted. Having grown impatient with Haupt's disregard for the military chain of command, Stanton went along with the plan. When Haupt again refused to sign his commission, Stanton fired him. Haupt left the army in September 1863.[41]

The rest of Haupt's life was a sad coda following his heroic assault on the Hoosac Mountain and his brilliant military record. Haupt had gone to the mountain a moderately wealthy man but never again enjoyed such financial security. After leaving the army, he invested badly in a stone quarry, a tourist resort, several farming ventures, an oil pipeline, an electric-light utility, and a powdered-milk company. For a time, he worked as general manager for the Northern Pacific Railroad but continued to lose money in poor investments. Haupt negotiated several financial settlements from Massachusetts but never enough to compensate him for the time and money he had wasted there. At one point, his wife, Anna, was forced to pawn her jewelry to pay the family's grocery bill. When he died in 1905 at eighty-seven, Haupt was still in debt and living on borrowed money. He is buried at Philadelphia's West Laurel Hill cemetery.[42]

So, what was Haupt's legacy in the Bay State? Without his oversized ambition and obsessive drive, the Hoosac Tunnel might well have been abandoned sometime in the mid-1850s. Too little progress had been made on the tunnel up to that time. Yet, when Haupt separated from the Troy & Greenfield Railroad in spring 1862, enough had been accomplished that the state could not give up the tunnel. While he had only bored 12 percent of the tunnel's eventual length, Haupt had completed most of the railroad track needed to connect the Hudson River with Greenfield, Massachusetts, and the Port of Boston beyond.

Moreover, Haupt had imbued townships like Greenfield, North Adams, and Williamstown with a level of commitment to the tunnel that had not existed before he arrived in the state. Without Haupt, the Hoosac Tunnel would have remained a shallow, abandoned cavity at the east portal and a wistful dream in neighboring communities.

Haupt failed for several reasons. First, he was hopelessly undercapitalized for a massive project like the Hoosac Tunnel. The cumbersome system of state-loan payments and the modest sums he could borrow from the private market were never enough to cover his expenses. Second, Haupt lacked the tunneling technology needed to accomplish such an undertaking. Such technology simply did not exist at the time. Although he clearly recognized its importance, he was not able to perfect and deploy a mechanical drill in time to help him at the tunnel. Third, Haupt was a builder but not a manager. His loose grasp of tunnel finances made him vulnerable to bad business partners and political enemies, both of whom accused him of fraud and crimped his resources. Another of Haupt's shortcomings as a manager was his inability to staff properly and delegate responsibility. This failure stretched him too thin on a project as complex as the tunnel. Finally, Haupt was a unique personality, possessing a commanding presence on the floor of the legislature and in the public square, but at times aloof and arrogant in one-on-one relationships. As a man of action, he was often impetuous and exercised poor judgment. His tendency to go behind the backs of his superiors, as happened with both Governor Andrew and Secretary of War Stanton, served him poorly. He was not what management experts today would call an "organization man." Still, for all his shortcomings and failures, Herman Haupt looms large in the story of the Hoosac Tunnel. Hercules, Prometheus, or Sisyphus. Take your pick. He was all of them.[43]

5

THE COMMONWEALTH INTERVENES

WHEN THE CIVIL WAR arrived in 1861, it seemed as though Massachusetts had invited it. The state was the epicenter of abolitionism. There was something about the place that "agitated the mass," Ralph Waldo Emerson wrote, "with some odious novelty or other." It was always stirring with "some new light," "some new doctrine," or "philanthropy." Emerson implied that this had to do with its "deep religious sentiment." Whether owing to the millennial zeal of its Puritan founders or more recent ethical notions, Massachusetts embraced the abolitionist cause with extraordinary passion. In 1831, William Lloyd Garrison began publishing his abolitionist newspaper *The Liberator* in Boston. Elite divines like Theodore Parker and Thomas Wentworth Higginson turned up the abolitionist heat with their sermons and tracts. Massachusetts congressmen Charles Sumner and John Quincy Adams took the fight to Washington; the former hectored slaveholders until one of them nearly beat him to death, and the latter eliminated the infamous "gag rule" barring antislavery petitions to the national legislature. During 1854, the city of Boston exploded in violence when federal marshals arrested fugitive slave Anthony Burns and shipped him back to the South. In 1859, the "secret six," a group made up of the city's most esteemed clergymen and philanthropists, helped finance John Brown's Harpers Ferry raid. With the outbreak of war, radical Republican John Andrew was the first northern governor to send troops to defend the nation's capital and the first to organize a black regiment. Massachusetts men from Salem to North Adams rushed to enlist. For the next five years, the war overshadowed the Hoosac Tunnel.[1]

And, yet, the war and the tunnel were oddly related. The war was an enormous state enterprise and, after Massachusetts took it over in early 1862, so was the tunnel. By then, supporters of the tunnel knew it was too large a project to be funded by anything less than state money. And why not, since the tunnel promised to bring new life to the state's economy? In the hothouse atmosphere of Massachusetts, so aptly described by Emerson, the war and the tunnel were easily conflated as triumphalist ventures. Unbounded confidence in the possibilities of moral and economic improvements ran deep in the state's collective psyche. Both the war to free America's enslaved population and the tunnel to access the bounty of the Golden West drew on this powerful impulse.

General Benjamin Butler, President Lincoln's tough military governor of New Orleans and later Massachusetts congressman and governor, articulated this relationship between the war and the tunnel when he visited the Hoosac Mountain during 1865. He described the tunnel as "the most gigantic undertaking in engineering the world has ever seen." He compared it to the war, emphasizing the sheer magnitude of both. Butler was confident that Massachusetts men, flush with victory from the battlefield, would soon defeat the mountain too. Alvah Crocker, who accompanied Butler on the visit, harbored similar sentiments. A staunch abolitionist, Crocker had fully supported the war. He had given so generously and worked so effectively to care for the state's wounded soldiers that Governor Andrew had placed him in an official post for that purpose. When Crocker heard that the Confederacy had surrendered at Appomattox, he marched to his paper mill and gathered his workers in a prayer of thanksgiving. After construction ceased on the tunnel in 1861, Crocker distanced himself from Herman Haupt but continued to proselytize about the project. In a speech the following year, he warned again that "Boston is fast losing its export trade" and was "becoming to New York what Salem is to Boston." The tunnel would "give the West a quick and cheap avenue to reach us," he promised. Crocker also began ingratiating himself with Governor Andrew. He declared that Andrew was "not to blame for discontinuance of the work." Crocker sat in the Massachusetts legislature during most of the war and, by the time it ended, had managed to secure the chairmanship of the joint committee over the Troy & Greenfield Railroad and the Hoosac Tunnel. With Crocker in that position, the tunnel stood a good chance of moving forward.[2]

But, first, Governor Andrew's three-man commission had to decide the tunnel's fate and, if it were to go forward, how it would be built. The commission was made up of three railroad presidents: Alexander Holmes of the Old Colony

Railroad; Samuel M. Felton of the Philadelphia, Wilmington & Baltimore; and John W. Brooks of the Michigan Central. Although the youngest at forty-two, Brooks was the most successful of the three and became the lead commissioner. Born in Stowe, Vermont, in 1819, he had studied civil engineering under Loammi Baldwin and served as superintendent of the Auburn & Rochester Railroad. He had helped Boston's wealthy financiers John Forbes and Nathaniel Thayer assemble their midwestern network of railroads and, with the wartime closing of the Mississippi to shipments of grain and other goods, made these men even richer. Given the challenges of the Hoosac Tunnel, the commissioners also enlisted the help of several well-respected engineers: James Laurie, a successful railroad builder with special expertise in bridges; Benjamin H. Latrobe, Jr., formerly chief engineer of the Baltimore & Ohio Railroad with experience boring forty-four tunnels and building three times that number of bridges during his career; and Charles S. Storrow, a brilliant mathematician and designer of both railroads and textile mills in Massachusetts and elsewhere. Storrow had graduated first in Harvard's class of 1829, studied engineering in Europe, and come out of retirement to serve on the tunnel project. While Laurie and Latrobe examined the condition of the Hoosac Tunnel, Storrow departed for Europe to observe the latest tunnel technology there, especially methods being used to dig the eight-mile Mont Cenis Tunnel through the Alps.[3]

At the end of February 1863, twenty months after work had ceased on the tunnel, Charles Storrow was back from Europe and the commissioners handed their report to Governor Andrew. After reading it, he declared that the tunnel was not only feasible but should be built by the state. The cost of the tunnel would be enormous, too large, the governor reasoned, for private industry to take on. State ownership offered the tunnel the kind of financial blood supply needed, not just to survive but to thrive. The tunnel was now estimated to cost $5,719,330. This amount included the $2 million originally authorized in the Loan Act of 1854, approximately half of which had already been spent. The higher cost of the tunnel was largely due to structural changes recommended by the commissioners. The most significant of these was the digging of a "central shaft," which would be sunk from the top of the Hoosac Mountain more than a thousand feet to the mid-point of the tunnel. The reasons for the central shaft were to establish two new headings, thereby speeding up work on the tunnel, to provide a source of ventilation for the tunnel, and to help ensure the proper alignment of the tunnel. There was considerable debate about the need for the central shaft. The Mount Cenis did not have one, and many of the experts Storrow interviewed in Europe did not believe it would be effective in ventilating the Hoosac Tunnel. However, Storrow argued that the slight incline of twenty-six feet per mile from the tunnel's two portals, designed to promote drainage and help in the removal of excavated rock, would force locomotive fumes to collect in the middle of the tunnel and could only be vented with the addition of a central shaft. Even though his data did not fully support it, Storrow was so adamant about the need for a central shaft that it was incorporated into the tunnel's new design. Another important recommendation was that the tunnel be enlarged to accommodate double tracks. This meant expanding its dimensions from 14 × 18 feet to 20½ × 24 feet. Critics of the tunnel had regularly mocked its diminutive size, and Francis Bird had graphically illustrated this inadequacy in his *Road to Ruin* pamphlet. Finally, the commissioners' report called for the tunnel's alignment to be straightened. Herman Haupt and Edward Serrell before him had aimed the two trajectories of the tunnel, one coming from the east portal and the other from the west portal, slightly off-center, in order to increase the likelihood of their meeting in the middle of the mountain. Known as "intersection digging," this was a standard but crude practice of tunnel boring at the time. Realigning the tunnel would be tricky. Achieving an absolutely straight path for the tunnel would require an elaborate surveying system across the top of the mountain, aligning the azimuth between the east and west portals with the work going on inside the mountain. The new azimuth could be accommodated fairly easily at the east portal. That part of the tunnel could be angled ever so slightly while it was being widened for double tracking. The new alignment also needed to pass through the west shaft, which no one wanted to dig again. However, the new alignment would not come out of the mountain at the existing west portal. It would have to be moved some five hundred feet east to correct its alignment and raised twenty feet to equalize its elevation with the east portal. To avoid the porridge-stone problem, the new west portal would be constructed in an open trench, arched over with several layers of bricks, then reburied. The tunnel would be arched as far into the west slope of the mountain as necessary to reach solid rock. A good source of clay had been discovered nearby. So, a brick factory was planned next to the west portal.[4]

These modifications alone raised the cost of the tunnel significantly. But there was more. The commissioners' report recommended the development of mechanical drills powered by compressed air (later referred to as "pneumatic" drills). Storrow had observed large drills of this type at work in the Mont Cenis Tunnel. They were eight feet in length and weighed close to four hundred pounds. Mounted in sets of nine on a carriage that moved on a railroad track, these drills had increased the rate of tunnel progress by two to three times that of hand drilling. Although he was oddly ambivalent about the use of mechanical drills in the Hoosac Tunnel—he did not think they would be less expensive than hand labor—Storrow was impressed by the use of compressed air as a power source for the drills. However, both Benjamin Latrobe and James Laurie were enthusiastic about the opportunity offered by the mechanical drill. "I have no question it will be used at the Hoosac Tunnel," Latrobe declared, "with such improvements as American ingenuity is apt to make on European inventions." As a consequence, the commissioners recommended that either steam engines be installed at the Hoosac Tunnel or a power dam be built across the Deerfield River to generate compressed air for mechanical drills. The latter option was decided upon and represented a major undertaking of its own. Construction of the Deerfield Dam would begin before a workable prototype of an American drill was in hand. While critics viewed this as financially incautious, the tunnelites were as convinced of the country's ability to invent a power drill as they were of their own eventual triumph over the Hoosac Mountain. They were not alone. In April 1863, the Massachusetts legislature approved a bill implementing the commissioners' new plan for the Hoosac Tunnel. It passed in the House by 131 to 47 and in the Senate by a margin of three to one.[5]

Nor was Massachusetts out of step with the federal government in this kind of legislation. Exhibiting the same assuredness with which the Bay State approved the new tunnel plan, the United States Congress committed to building a transcontinental railroad from Council Bluffs, Iowa, to Sacramento, California. The Pacific Railway Act of 1862 was the federal government's first corporate charter since 1791, when it had created the Bank of the United States. This generous piece of legislation and a similar one two years later lent the Central Pacific Railroad and the Union Pacific Railroad $75 million in interest-free bonds and provided them land grants the size of entire states. Yet, the transcontinental railroad served no existing commerce and traversed mostly empty space. Though this munificent legislation had broad support in Congress, some there pointed out its absurdity. "There is no travel from the Pacific coast to justify it," F. A. Pike of Maine protested. "Here are 1800 miles of railroad through *an uninhabited country.*" Like the Hoosac Tunnel, the transcontinental railroad intended to capitalize on the perceived bounty of the West. So, too, it became conflated with the war effort. Many believed it would tie the western states and territories closer to the Union. Congressman Timothy Phelps of California called it a "military necessity." Such was the spirit of the times in both the Congress and the Massachusetts State House.[6]

The race for the West had become more intense as various states completed railroad routes from the Northeast and Mid-Atlantic regions to the Midwest and beyond. The Erie Railroad had been the first in 1851 and was soon followed by the New York Central, the Pennsylvania Railroad, and the Baltimore & Ohio. By the mid-1850s, four competing railroads lines had spanned the trans-Allegheny region and circumvented the Erie Canal with connections to the Ohio River and the Great Lakes. By the second half of the decade, other rail systems had reached the Mississippi River. The Chicago, Burlington & Quincy as well as the Rock Island were there by 1856, the latter being the first to build a bridge over the great river. As these railroads built westward, they benefited from telegraph lines, in wide use by the late 1850s, for better scheduling of their freight and passenger traffic. As the country underwent a "railroad revolution," joining eastern seaboard cities with far-flung western towns in reciprocal trade, Massachusetts was still struggling to build a viable rail line from Boston to the Hudson River.[7]

In June 1863, Commissioner Brooks announced the appointment of Thomas Doane as the new chief engineer of the Hoosac Tunnel. Born in 1821 in Orleans, Massachusetts, on Cape Cod, Doane had attended the Phillips Academy before joining Samuel Felton's engineering firm in Charlestown. He had been involved in the construction of Alvah Crocker's Fitchburg Railroad as one of his earliest assignments and later served as chief engineer of the Vermont Central Railroad. By the time he packed his topographical maps and surveying transit for North Adams, Doane was at the top of his field as a railroad engineer and regarded as a precise and exacting professional by those who knew him. Coming from an engineering firm rather than a powerful corporate culture like the Pennsylvania Railroad, Doane cut a lower profile than Herman Haupt had, struck people as less ego-driven than his predecessor, and dedicated himself to simply doing good work for his clients. Nonetheless, in implementing the new tunnel plan, Doane would advance mining technology further than he could have imagined as he departed for the Hoosac Mountain.[8]

Restarting work on the Hoosac Tunnel and the adjoining railroad tracks seemed to take forever. Haupt's roadbeds had eroded terribly and the bridge over the Green River was still in a state of collapse. Because of the tunnel's realignment, the west portal had to be abandoned and begun again at a new location. Nor could the tunnel from the east side of the mountain be penetrated further until it had been widened and realigned. The west shaft was filled with water. Furthermore, the war made finding labor difficult. Nonetheless, recruiters fanned out to Boston, New York City, and parts of Canada. Mostly, they targeted poor, immigrant neighborhoods. By the end of 1864, there were six hundred men working at the Hoosac Mountain. Doane constructed better living quarters for his workers and their families, believing that "a better class of shanties" would attract "a better class of workers." There were thirty-six dwellings each at the east and west portal, nine at the central shaft, and thirteen at the Deerfield Dam. Doane also built several company stores. A large one above the east portal housed a school for sixty to seventy children on its second floor. Religious services were held there on Sundays. "Men of family are more desirable than single men," Doane observed, because they would work for "a considerably lower price." Wages for common laborers were $1.50 per eleven-hour day and for miners $1.75 per eight-hour day. Out of these wages, workers had to pay rent for their shanties and food from company stores. Food was provided at cost, and each worker carried a kind of "credit card" showing his accumulated wages and purchases. Doane kept careful records for each three-man drill team, consisting of a holder and two strikers. With closer record keeping, drill teams' average depth drilled per day rose from six feet two inches to ten feet and then thirteen feet per day. Only the better drill teams were retained. Any worker "induced by evil to strike for higher wages or any other cause" was immediately let go and lost any accumulated wages. Doane was never able to establish a system to measure the work of common laborers because their tasks were too varied. Nor was he able to control the problem of liquor consumption, which caused "much trouble and interfered with the orderly conduct of work." With workers laboring under unsafe conditions with low pay and winter temperatures as low as ten degrees, recourse to alcohol was understandable.[9]

So, who were Doane's laborers at the Hoosac Tunnel? Where did they come from? What were their lives like at the tunnel? Regarding their countries of origin, exact breakdowns are difficult to come by and they changed over time. However, as noted before, accident reports in local newspapers and vital records in townships near the tunnel provide a good sampling of tunnel workers and often include their names, ages, countries of origin, and family status. It appears that some 40 to 50 percent of tunnel laborers were Irish. Between 10 and 15 percent were either French Canadian, English, or Scottish. The balance were either local residents or unidentified as to origin. The Irish came to the tunnel mainly from the Boston area. Most had been in and around that city for several years. When they could find work at all, they had labored in textile factories or on construction sites where they had been poorly paid. Wages at the tunnel were somewhat higher and living expenses lower. Some of the Englishmen came from Cornwall and brought copper mining experience with them. A few had served in the Crimean War and were hardened by the severe travails of that conflict. French Canadians arrived by railroad from Quebec Provence, which suffered periodic depressions and high unemployment. Some third or more of tunnel workers had families. In terms of age, approximately 40 percent of tunnel workers were in their twenties and about the same proportion in their thirties. The balance were teens or over forty. The physical demands of tunnel work did not attract many older workers. All tunnel workers came from desperate circumstances elsewhere. No one came to work at the tunnel if they didn't have to.[10]

In spite of Doane's efforts to improve living conditions at the tunnel, life for workers and their families was harsh and dangerous. Most shanties were squalid and too small for entire families. Clustered in close proximity, worker villages were inherently unhealthy. Outbreaks of measles were a constant problem and infected dozens of children at a time. Food was always scarce, especially if the family breadwinner was a heavy drinker. During Berkshire winters, temperatures dropped well below freezing. Women were forced to forage for firewood with their children. Worst of all, the risk of fatal or maiming accidents hung over the community like a specter. There were so many ways to die. Although the actual number will never be known, recent research has verified that at least 135 men died at the tunnel. Another sixty or so were so badly injured they could not work again and may have died from their injuries shortly after leaving the tunnel. The most common accidents were caused by explosives inside the tunnel (mostly fuses hanging fire), followed by falling rock in the tunnel and falls by miners themselves from ladders and scaffoldings. Anywhere near the tunnel was dangerous. Carts overturned crushing their drivers; men froze to death going to and from their work sites; and others died manufacturing explosives or transporting them to the miners. The central shaft was by far the most hazardous place to work at the tunnel. Men fell into it, drowned in its wet bottom, or were skewered by falling drills. It became known as the "bloody pit."[11]

Controversial from the start, Doane's greatest challenge was the central shaft. To begin with, he was forced to move its location slightly closer to the new west portal because the Cold River on top of the mountain flowed over the exact middle point of the tunnel's intended path below. Even with this adjustment, the area around the central shaft was wet and prone to flooding. Consequently, a cement wall ten feet thick and thirty feet deep had to be constructed to reinforce the opening of the central shaft and a drainage system was installed around it. The opening itself was in the shape of a twenty-seven- by fifteen-foot ellipsis, its long side running along the line of the tunnel. This would allow room at each end of the ellipsis for a large bucket hoist, one per heading once the shaft reached tunnel grade. During the digging of the shaft, blasted rock would be raised in the buckets and miners sent up and down in them. The center of the shaft would be occupied with a stack of sixty-two round platforms connected by ladders, pipes to raise pumped-out water, and hoses to deliver fresh air down to the work site. A hundred-horsepower steam engine would hoist the two buckets, and smaller engines would drive the water pumps and ventilation system. Doane purchased 250 acres of woodland near the central shaft to provide fuel for the steam engines. Miners working on the central shaft hit solid rock at twenty-five feet and, by the end of 1864, had penetrated down to seventy-five feet. The full 1,028-foot depth of the central shaft would not reach the grade of the tunnel until summer 1870. It would be the deepest man-made hole on earth.[12]

Doane selected a site three-quarters of a mile up the Deerfield River from the east portal for his power dam. While the river lacked a solid bottom most of the way, there was at least a rock shelf jutting out from the shoreline there on to which to attach the dam's footing. The dam consisted of log "cribs" tied together with heavy steel rods and filled with stones. This so-called crib dam extended 250 feet across the river and had two aprons, an upriver one seventeen feet high to hold the water back and directly behind it a lower one which descended to the natural riverbed. This stepped structure allowed the river to flow over the dam during rainy periods without damaging it. To prevent undermining by the river's flow, a cement wall was built under the lower apron six feet into the riverbed. The pond behind the dam extended a mile upriver. On the west side of the dam, a canal carried water to a compressor building near the east portal. A fall of thirty feet was expected to provide enough hydraulic power to drive the turbine wheels there and in turn the air compressor. The compressed air was estimated to be at least sixty pounds per inch, enough to not only power the mechanical drills at the east portal but be transported over the mountain by pipes to the central shaft and west portal without a significant loss of pressure. All of this was highly theoretical and still awaited a workable mechanical drill. It was also very costly. By the end of 1864, the dam and its attendant buildings and equipment had cost the state $252,917, much more than the $162,565 spent on the tunnel modifications.[13]

The new surveying system to straighten the tunnel's alignment was the most tedious aspect of Doane's work. His calculations had to be so precise that miners boring from east and west would eventually meet face to face somewhere in the middle of the Hoosac Mountain. Given the length of the tunnel and the chance for error, this was a tremendous technical challenge. To start with, Doane cleared a twenty-foot-wide swath over the Hoosac Mountain from the east portal to the new west portal. A transit team ensured that survey stakes set along the swath formed a perfectly straight line. Taking a page out of the Mont Cenis learning, Doane then built four "aligning towers," two on the east and west peaks of the Hoosac Mountain, where the cleared swath crossed over them, and two on nearby hills, one directly east of the east portal and the other directly west of the new west portal. Each aligning tower consisted of an approximately twenty-square-foot stone building with wooden roof, topped by a six-foot red and white pole for easy sighting from a distance. Smaller "sighting monuments" were erected just outside both portals and on top of the central shaft. Using the latest high-powered transit scopes available to him, Doane established the exact azimuth he wanted across the mountain and visually aligned it with the two portals and central shaft. Inside the mountain, he suspended pointed plumb bobs from wooden pegs driven into the tunnel's ceiling and aligned these plumb bobs with the azimuth he had established over the mountain. To do this, he used the light visible at the east and west portals and, more precisely, the sighting monuments just outside them. In this way, Doane ensured that the line of the tunnel matched the azimuth he had struck over the mountain.[14]

Wandering around the east portal, Consulting Engineer James Laurie had a good laugh looking at the rusting hulk of Herman Haupt's giant boring machine. (It sat not far from the earlier Excavator.) He noted its high cost and called it a "monument to misapplied ingenuity." Still, both Laurie and Tunnel Commissioner John Brooks believed the success or failure of the Hoosac Tunnel depended on the development of a viable mechanical drill. Without it, the slow rate of progress with hand drilling would exhaust the patience of the public and spell the end of the tunnel. As a consequence, the commissioners placed a high priority on the development of a mechanical drill in their report to Governor Andrew. It remains a mystery why four Mont Cenis drills, supposedly ordered during Storrow's visit there, never arrived at the Hoosac Mountain. It may have had to do with Storrow's ambivalence regarding the need for mechanical drills or, perhaps, Brooks's interest in developing his own drill at the state's expense.[15]

In their December 1864 progress report, the commissioners mentioned that they had acquired plans for several mechanical drills. These probably included plans for the Mont Cenis drill that its designer Germain Sommeiller had given to Storrow, as well as for a number of American drills that had made it to the prototype stage but had not been field tested. Plans for the American drills were available from the U.S. Patent Office. Among the most promising were those designed by Philadelphians Jonathon Couch and Joseph Fowle during the late 1840s. These were steam-powered drills that consisted of a hollow cylinder containing a piston attached to a lance-like drill bit. A gearing system rotated the drill bit after several strikes, roughly imitating the action of the holders and strikers in hand drilling. This basic design concept would serve as the model for most of the drills to follow. Still, important modifications would have to be made before these designs could be expected to hold up under the brutal, jarring impact of boring through solid rock. The development and testing of these design modifications would be tedious, expensive, and prone to failure. Unfortunately, neither Couch nor Fowle had the necessary capital to pay for such development and testing. Someone who did was William Harsen of Brooklyn, New York. However, Harsen's drill failed in two different tests before the tunnel commissioners. It had literally flown apart attempting to penetrate the adamantine hardness of Hoosac rock.[16]

Still, Brooks sensed that American inventors were tantalizingly close. What is more, he believed he could come up with the winning design first and profit from it personally. He formed a partnership with the Putnam Machine Company in Fitchburg, Massachusetts, whose brilliant mechanic Charles Burleigh seemed to understand the kind of drill needed at the tunnel. It had to smash into Hoosac rock with enough force to drive a two-inch bore several feet deep, the depth required to place an explosive charge. Furthermore, the drill had to hold up to this violent engagement with only occasional breakdowns and parts replacement. It also needed to be light enough to be removed from the tunnel heading before blasting. Finally, the ideal drill had to be driven by a power source friendly to the tunnel environment, preferably compressed air. The Putnam Machine Company rented shop space from Alvah Crocker and had a long-standing relationship with him, so Crocker was aware that Brooks was lavishly funding Burleigh's drill development with state money, all the while taking patents out in his own name. Crocker was so wealthy by this time that he didn't care. Since Crocker's first priority was the Hoosac Tunnel, he was willing to look the other way if Brooks and Burleigh could come up with a compressed-air drill to speed up work on it.[17]

Though he had abandoned the Hoosac Tunnel in 1861, Herman Haupt also coveted the prize Brooks was after. Haupt continued to work with Stuart Gwynn, his drill inventor while at the tunnel, throughout the Civil War. Both were determined to perfect the prototype they had developed but never actually field tested. Haupt even apprenticed his oldest son, Jacob, to a Philadelphia machine shop and, once Jacob came of age, set him up in his own shop to work on the drill. Even though their drill was based on the earlier designs of Couch and Fowle, Haupt and Gwynn were able to secure several patents of their own during the 1860s. Among their modifications were a ratchet to rotate the drill bit and a gripper to regulate its feed into the rock. The more they worked on their drill, the more the two men came to believe the Couch-Fowle drills were their own inventions. In 1862, Haupt wrote Commissioners Felton and Brooks asking that their drill be adopted at the Hoosac Tunnel, and, a year later, he formally made the Massachusetts legislature aware of his and Gwynn's pending patents. During the summer of 1865, Haupt, his son Jacob, and Gwynn began testing their drills and ancillary equipment at a coal mine in Wiconisco, Pennsylvania, southwest of Harrisburg. The test was a disaster. The drills continually broke down, the drill carriages proved too cumbersome to remove before blasting, and the steam hoses to the drills leaked badly. Several times, the miners were overcome with noxious fumes. After several months, Haupt ran out of money and moved his equipment back to Jacob's machine shop for additional modifications.[18]

Meanwhile, Brooks kept a wary eye on Haupt. Though Brooks could outspend his rival with unlimited access to the state purse, he worried about the threat of patent infringement from Haupt and the length of time Burleigh was taking to produce a workable drill. To address the first problem, Brooks secretly purchased Couch's original patent on the piston-cylinder drill, thereby protecting any subsequent design based on that concept. To address the second problem, Brooks invited members of the Massachusetts legislature to a demonstration of the Burleigh drill in Fitchburg at the end of 1865. During the test, the Burleigh drill bored six holes totaling 113 inches, at an average of three inches per minute, in rock brought from the Hoosac Mountain. Even Haupt, who attended the demonstration, was impressed. Not knowing Brooks had purchased the Couch patent, Haupt warned those present that the drills infringed on his and Gwynn's patents. By the middle of the next year, Brooks had forty of the Burleigh drills at the east portal of the tunnel. Once again, however, experience showed that small-scale tests of mechanical drills were poor predictors of their performance at actual work sites. After five days at the tunnel, the drills began to fall apart. Doane counted over a thousand failures in the second half of 1866. Fortunately, Burleigh acknowledged these failures quickly and, by the end of the year, had redesigned the entire drill. He went back to Fowle's original idea of attaching the drill bit to the piston for extra stability. This allowed the bit and piston to rotate together, an idea William

Harsen had advocated. Burleigh's real breakthrough, however, was directing the exhaust air from the drill in such a way that it cushioned the bit and piston assembly after each strike into the rock. If there was one innovation that made the drill successful, it was this idea of using redirected compressed air to create this cushioning effect. Burleigh's new drills not only held up well but doubled the rate of tunnel penetration versus hand drilling. When fully developed, the new drill exerted fifty pounds per square inch and struck into rock 250 times per minute. Whereas the headings had been advanced at the rate of fifty-five feet per month before, the new Burleigh drills achieved 116 feet. Part of this success was also due to a track-mounted drill carriage Doane had designed, that could be quickly removed from the tunnel heading before blasting and also allowed the miners to maneuver the 372-pound drills easily when choosing where to start new drill holes. Similarly, a pair of air compressors were developed by Doane and a North Adams firm to provide power for the mechanical drills and cleanse the tunnel air after blasting. Key here was the development of a cold-water injection system to cool down the extremely high temperature of the compressed air so that it could be safely conducted via rubber hoses from the compressors to the drills inside the tunnel. By the end of 1866, miners at the east portal had laid down their sledge hammers and star drills for the last time and were becoming familiar with their new mechanical drills. The "Burleighs," as they were affectionately called, would revolutionize drilling technology and over time become ubiquitous at mining and tunneling sites across the country.[19]

Cracking the code on mechanical drilling would ensure much better progress on the Hoosac Tunnel. However, it would come too late to help Brooks, who suffered a debilitating stroke while testifying before the Massachusetts Senate in May 1866. Fulfilling his duties as president of the Michigan Central Railroad during the Civil War and serving as lead commissioner for the tunnel had physically destroyed him. Although he would later recover, Brooks was forced to resign as a commissioner and would never again be involved with the tunnel. The new governor, Alexander H. Bullock, who took over from Andrew at the beginning of the year, appointed Alvah Crocker as the new lead commissioner. James Shute, a political wire-puller from Somerville, Massachusetts, replaced Commissioner Samuel Felton, who had also suffered a stroke. Except for the arrival of the Burleigh drills, conditions at the tunnel had continued to deteriorate. Once again, work inside the west shaft had to be shut down because of flooding. During late 1865, miners went on strike and burned down several buildings near the west shaft. The number of accidents mounted at an alarming rate as miners penetrated further into the east portal, dug deeper down the central shaft, and struggled to shore up the west portal. In 1866 alone, fourteen miners were killed or seriously injured. Hung-fire blasting accidents in the east portal, falling equipment in the central shaft, and cave-ins at the west portal all claimed victims. Especially worrisome to Crocker, the cost of the tunnel soared in an environment of post–Civil War inflation. Budget additions for the new drills, air compressors, drainage pumps, and brick arching pushed the tunnel's cost higher. Brooks had collapsed after securing $900,000 in new funding from the state, on top of the $2,818,589 already spent.[20]

Adding to the pressure on the commissioners, Francis Bird published another of his splenetic pamphlets during the summer of 1865. *The Hoosac Tunnel: Our Financial Maelstrom* took aim at Brooks personally and pointed out the seeming futility of the tunnel. Bird characterized Brooks as not only "arrogant, insolent, domineering" but also a "lacky" of railroad magnate John Forbes. Brooks's only interests in the tunnel, Bird argued, were to link the Troy & Greenfield Railroad with Forbes's Michigan Central line and profit personally from mechanical drill patents paid for with state money. Most serious, however, Brooks had "lamentably failed in planning and organization" at the Hoosac Tunnel. Among his chief follies was the Deerfield Dam, which would likely cost over $275,000 and supply insufficient hydraulic power for the tunnel. The central shaft was another major mistake. Its unprecedented depth would take years to penetrate and reach the tunnel's grade too late to open productive headings. Bird quoted expert testimony that the central shaft would be ineffective in ventilating the tunnel. So, too, Bird estimated that protecting the new west portal from flooding and porridge stone could run into millions of dollars. And what, after all, had Brooks accomplished? He had penetrated 1,145 feet into the Hoosac Mountain, compared to 2,439 feet accomplished by Herman Haupt at a much lower cost. With 21,000 feet of tunnel yet to go (of its total 25,000 feet), who knew what the tunnel would ultimately cost? Conservatively, Bird estimated it could cost over $10 million. Bird concluded his attack with the ironic plea—"Oh, for one year of Herman Haupt!"[21]

Bird recommended abandoning the tunnel at once, writing it off as a total loss, and devoting future state aid to more practical projects. His refrain was picked up by anti-tunnel newspapers across the state and legislators opposed to the tunnel. The *Boston Courier* noted Brooks's poor performance driving the tunnel versus Haupt before him. In the legislature, anti-tunnelites accused the commissioners of "groping their way without any sufficient guidance from experience elsewhere." These anti-tunnel legislators dismissed most components of the commissioners' tunnel strategy—from the Deerfield Dam to the central shaft—as "entirely chimeral."[22]

And, yet, Alvah Crocker was able to ram another $500,000 in tunnel funding through the Massachusetts legislature in the wake of Bird's diatribe. How was this possible? Partly, it was due to the fact that Massachusetts had become a one-party state and was preoccupied with issues at the national level. The Democrats, who might have opposed such lavish spending, had been demolished in the debate over slavery. With the ascendancy of Andrew to the governorship, the so-called Radical Republicans dominated state politics and devoted most of their energy to winning the war and reconstructing the South. This meant that important state issues like liquor laws, workingmen's hours, and control of public spending went largely unaddressed. Importantly, this also meant that special-interest groups and potent local initiatives were able to operate stealthily beneath the party's war and reconstruction platform. Crocker and his tunnelites deployed their lobbying efforts under the "cover" of the party's southern agenda. Bird and his allies were less adept at these tactics and much less flexible in their approach to government. They were loath to dilute their stance on the war and reconstruction with state and local issues. The situation was made even easier for the tunnelites because Republicans at their core had always believed in fiscal spending behind public services and internal improvements. In 1865, Governor Andrew approved a hugely expensive bill establishing a state-wide constabulary (the nation's first police force). Legislation to improve Boston's harbor soon followed. An expensive war had habituated the party to this kind of spending. Fiscal restraint would have to wait for the emergence of Liberal Republicans and recovery of the Democrats several years away. Ironically, both Crocker and Bird were deeply committed Radical Republicans, and both believed in internal improvements. What they were arguing about was *how* and *where* to spend the state's money. They each embodied their respective localism and particular corporate interests. In the existing political environment, Crocker proved the superior lobbyist and more nimble State House tactician.[23]

Also, Crocker had the more inspiring vision when it came to internal improvements. The Hoosac Tunnel was gaining, not losing, popular support. For those who believed in it, the tunnel promised an economic millennium for Massachusetts. The tunnel fed off the twin notions of immense western bounty and the ability of the state's railroads to deliver that bounty to the wharves of Boston. The Conference of Western Trade Representatives, held in Boston during the summer of 1865, showed how desperately the public craved western access and its potential prosperity. Boston newspapers shamelessly ingratiated themselves with their western visitors and flaunted the alleged advantages of their city for the export of western products. Bostonians, the newspapers observed, were "greatly pleased with the solid men of the western commercial centers." The newspapers urged their hosts to show these men the state's "unsurpassed" railroad system and a harbor which was "naturally the finest." The western visitors were treated to lavish dinners, special tours of Boston's harbor and warehouse facilities, and bombastic speeches laden with postwar patriotism and exaggerated visions of mutual prosperity. The dust of eastern and western men had mingled together on the battlefields. They were a "band of brothers." "With one arm it reaches out to the prairies," one Boston booster described his city's commerce, "and with the other it lays the Eastern hemisphere under tribute for the wants of itself and the far interior." Reciprocating such sentiments after many toasts, a Chicago merchant promised that "if you reach out your iron arm there, with sufficient cars to transport grain, you can land our grain on the wharves of Boston." Boston, the Chicagoan continued, could export that grain "in successful competition with any other point on the Atlantic coast." It was music to Bay State ears. As they said goodbye to their western visitors, the Bostonians promised a "speedy completion" of the Hoosac Tunnel.[24]

Nonetheless, Crocker understood that he could not take popular support of the tunnel for granted. While some of Bird's criticism of the commissioners had been unfair, Crocker recognized that the tunnel was too costly and demonstrating too little progress in the public's simple view. Crocker's approach going forward was obvious: cut costs wherever possible, capitalize on new drilling technology, and drive tunnel penetration faster. To achieve this, Crocker needed tighter control of the tunnel project. He quickly alienated Alexander Holmes, the last of the original three commissioners, who resigned in disgust at Crocker's heavy-handedness. Crocker replaced him with ex-railroad man and Lexington politician Charles Hudson. In spite of Hudson's prior connection to the Western Railroad, Crocker believed he could trust him. Soon, however, Hudson and James Shute were calling attention to Crocker's penny-pinching and autocratic management style. In early 1867, Governor Bullock, who seemed to have Crocker's back politically, replaced both men with Berkshire County senator Samuel Bowerman and Judge Tappan Wentworth of Lowell, both of whom were strong tunnel supporters. Bowerman was an arch enemy of Francis Bird and had done battle with him for years. Crocker now had better control over the tunnel but not total control. With Crocker's appointment as lead commissioner, the Massachusetts legislature had stipulated that Benjamin Latrobe be brought back to the tunnel as consulting engineer, reporting directly to Governor Bullock. Furthermore, Thomas Doane was still chief engineer, reporting to Crocker. Many, including supporters of the tunnel, blamed Doane for mistakes made during the Brooks era.[25]

For now, however, Crocker and Doane cooperated on fleshing out the technology needed to accelerate work on the tunnel. More specifically, Doane had been experimenting with simultaneous versus sequential blasting. He had shown that igniting multiple blast holes at the same time ripped away much more rock using the same amount of explosive as igniting them sequentially. Still, while relying on a variety of domestic and imported ignition devices, Doane had experienced continuing supply, cost, and reliability problems. Reliability was a matter of life and death for his miners. Eventually, Doane turned to Charles and Isaac Browne of North Adams, who had developed an electrically ignited fuse consisting of fulminate of mercury combined with several less expensive explosive materials. The fuse was essentially a wooden capsule, an inch and a quarter long, with two copper wires barely separated inside a highly sensitive fulminate core. When ignited, a spark jumped between the wires and set off the fulminate and the explosives next to it. The Browne brothers' fuse proved more cost-effective and dependable than competitive offerings. Soon, thousands of their fuses were being used at the tunnel. While the brothers would profit handsomely from their tunnel business and a subsequent patent, Charles Browne would later be blinded by an accident in their manufacturing facility.[26]

Such were the hazards of working with highly volatile ignition devices and explosives. Nitroglycerin took these risks to a new level. Even the Italian chemist Ascanio Sobrero, who first synthesized nitroglycerin in 1847, warned against using it as an explosive. Nonetheless, nitroglycerin's tremendous power made it irresistible to miners and tunnel builders. Swedish chemist and entrepreneur Alfred Nobel partially tamed nitroglycerin and made it commercially available by mixing it with diatomaceous earth and marketing it as "dynamite." In the process, he lost his brother, numerous coworkers, and at least two factories in massive, accidental explosions. During the summer of 1866, Colonel Taliaferro P. Schaffner, Nobel's representative in the United States, visited the tunnel and demonstrated for Doane and Crocker the power of dynamite versus black powder. It appears that these tests were successful. However, procuring and transporting the new explosive was problematic. In 1867, Nobel shipped three crates of his product to San Francisco for use in the Central Pacific Railroad's Summit Tunnel. One of the crates blew up in the city's Wells Fargo office, totally demolishing the building and killing fifteen people. Many cities in Europe and the United States banned shipments of nitroglycerin. Nobel eventually exited the business, but others like Professor George Mordey Mowbray took up the slack. A British chemist with a sense of adventure, Mowbray had gone to the newly discovered oil fields of Titusville, Pennsylvania, to explore nitroglycerin's potential. Mowbray's product was a highly purified, clear liquid he called tri-nitroglycerin. After seeing Mowbray's advertisement in *Scientific American*, Crocker invited him to the Hoosac Tunnel. During the summer of 1867, Mowbray began producing his product in an "acid house" at a safe distance from the west portal. Experiments at the tunnel showed that tri-nitroglycerin was eight times more powerful than black powder. Purely by accident, Mowbray discovered that tri-nitroglycerin could be safely transported when frozen. A wagon carrying tri-nitroglycerin had turned over in the snow and Mowbray's concoction was frozen solid. Efforts to ignite it showed it to be benign as long as it was frozen. Henceforth, tri-nitroglycerin would be delivered to miners at the central shaft and east portal in a frozen state and then thawed out before being ignited. In spite of these precautions, Mowbray's production facility would blow up a few years later when an errant spark from a stove set off four hundred pounds of finished product. Mowbray's foreman John Velsor would be blown to bits. Shreds of his flesh would be collected from

nearby bushes for burial in a sack. Even with these risks, tri-nitroglycerin became the explosive of choice at the Hoosac Tunnel due to its unrivaled rock-
shattering force. Almost half a million pounds of it would be used in the
tunnel. With Burleigh's pneumatic drill, the Browne brothers' fulminate-of-
mercury fuses, and Mowbray's tri-nitroglycerin—mining technology had entered the modern era at the Hoosac Tunnel.[27]

Other aspects of work at the tunnel had mixed effect. On the positive side,
Doane seemed close to gaining control over the problematic west portal. Rec-
ognizing the futility of trying to tunnel through the oozing mass of porridge
stone and constant inundations of water from inside the mountain, he dug a
thirty-foot-wide, open trench several hundred feet into the side of the mountain. He reinforced the sides of the trench with timbers and wooden planks,
installing drains, or "adits," at the bottom of the trench to sluice water off into
a nearby basin. Doane then hired B. N. Farren, a contractor from Doylestown,
Pennsylvania, who had done grading work for Herman Haupt a decade earlier,
to construct a giant brick tube inside the open trench. The sides of Farren's
tube would be two feet thick and consist of eight layers of brick and mortar.
The lower half of the tube was referred to as the "invert" and was built first. It
contained the rail bed on top and an elaborate system of drains underneath.
The upper half of the tube was built with the help of wooden "galleries" to provide protection from the porridge stone and allow masons to lay up the brick
work without fear of landslides. With work beginning during spring 1866, Farren completed the first 174 feet of the tube by year end. Earth was then backfilled around the tube. After that, the tube entered the mountain itself, using
even heavier timbers in the galleries and installing more capacious drains behind them to keep the work site dry. One million board feet of hemlock timber
would be used to construct the galleries. The drains Farren designed into the
brick tube were critical. The flow of water was estimated at over seven hundred
gallons per minute. Farren's massive tube would eventually extend 931 feet into
the mountain and require over twenty million bricks. Two large kilns located
near the west portal worked nonstop for three years keeping Farren supplied
with them. The work was painstakingly slow and not completed until early
1869. As Francis Bird had predicted, it was also very costly. Without it, however, there would be no tunnel.[28]

Other aspects of the tunnel proceeded less successfully. The Deerfield Dam proved disappointing. Bird had been right about its inadequacy. While a surplus of water flowed over the dam in early spring, flows diminished in summer and were insufficient to power the air compressors. Furthermore, the river froze in winter, and the air compressors ceased to operate entirely. As a result, steam-powered compressors had to be installed at the west portal and west shaft during 1866 and, a year later, at the east portal too. Additionally, increased flooding at both headings of the west shaft required larger pumps than those originally installed there. A new shaft was sunk 264 feet to the west to try to drain the west shaft but was only partially successful. At times, the water level in the west shaft rose to nine feet. As the costs of this new equipment and construction put pressure on Crocker's budget, his penny-pinching became more erratic. During the first week of 1867, he fired Thomas Doane. The precipitating issue between the two men was probably the need for larger pumps at the west shaft. Doane had recommended spending $50,000 for the larger pumps, but Crocker stonewalled until the west shaft was completely flooded and work ceased there. Doane did not go peacefully. He wrote an open letter accusing Crocker of purposely sabotaging the west shaft in order to transfer manpower to the east portal. He also claimed that Crocker required many of his laborers to work on the Sabbath. These charges were never proven, and Doane moved on to become the Burlington Railroad's chief engineer. Crocker replaced Doane with Charles Manning, an experienced tunnel builder from the Baltimore & Ohio Railroad, probably recommended by Latrobe, himself a veteran of that line. After four months of being ignored by Crocker and cut out of decision making, Manning resigned in frustration. Crocker then turned to W. P. Granger, a novice engineer hired at a quarter of Doane's salary. In his year-end report, Crocker wrote that "in all cases where the commissioners believe the salary paid was of more value to the State than the services rendered, men have been discharged."[29]

Out of this management chaos, one good thing emerged that probably saved the Hoosac Tunnel in the long run. The performance of B. N. Farren in driving the arching inside the west portal impressed both Crocker and Latrobe. Although the two men rarely communicated and did not agree on much, both saw the so-called contract system as the best way to organize work on the tunnel going forward. According to that system, work would be contracted out to several firms to complete various parts of the tunnel at an agreed price with strict deadlines. In other words, the work force at the tunnel would no longer be managed directly by the commissioners, their chief engineer, and his staff. B. N. Farren would retain responsibility for work at the west portal and be given the track from the east portal down the Deerfield Valley, including rebuilding Haupt's collapsed bridge over the Green River. The Harrisburg firm of Dull, Gowan & White would take over work at the east portal and the central shaft. An important stipulation was that the firm would have the use of the state's equipment at both locations. Dull, Gowan & White would have to return this equipment to the state in good condition when their work was completed. Crocker signed the contract with the firm in the summer of 1867, just months before disaster struck at the central shaft.[30]

6

DISASTER AND RECKONING

AS OCTOBER 20, 1867, dawned, a mist covered the valley around the central shaft. By noon the mist had lifted, and the seasonal chill had surrendered to a fragile warmth. To the east and west, sugar maples flashed red against the pine-green hillsides. The sheds and shanties around the central shaft came alive. The scent of bread and coffee filled the air. The clatter of pots and pans mixed with the throb of motors and rasp of tools being sharpened. The thirteen miners waiting in front of the central shaft building were dressed in dark, baggy clothes meant to keep them warm down below. Some smoked pipes and admired the foliage on the distant hills. Since it was Saturday, some of the miners, especially those with families, may have had plans for the Sabbath. Around one o'clock in the afternoon, the signal to descend into the central shaft came with the dull thud of an explosion below. The foreman and his assistant, who had ignited the blast, emerged from the building. One of the miners, James Fitzgerald, was the foreman's brother. They may have exchanged pleasantries before Fitzgerald and a dozen of his colleagues entered the building, climbed into the two large iron buckets, and began their descent into the central shaft. Once at the bottom, they would begin loading debris from the blast into the buckets.[1]

By now, the central shaft had penetrated to 580 feet and was crowded with structures and equipment to support mining at that depth. It was anticipated that, once the shaft reached tunnel grade (another 448 feet), the two iron buckets would be used to haul up debris from the shaft's east and west headings. Consequently, passageways for the buckets were provided at each end of the shaft's 27 × 15 foot elliptical opening. In the space between these passageways, the miners had constructed wooden platforms every twelve to twenty feet and secured them to the shaft's rock walls. Ladders connected these platforms all the way up and down the shaft but did not span the distance between the lowest platform and the worksite at the bottom. At that level, blasting and flying rock would have destroyed any ladder. This was one reason the miners chose to ride down the shaft in the buckets and not use the ladders. Speed was another consideration, as the shaft went deeper. There was also the danger of falling from the wet, slippery platforms and rickety ladders. Two months earlier, two miners had died that way. Since hand drilling was still used in the central shaft, the miners had developed the bad habit of storing their star drills and hammers on the platforms. One miner had been impaled by a falling drill the previous spring. Fastened to the platforms were a series of rubber hoses that served a pair of hydraulic pumps and a forced air ventilator housed in the building at the top of the shaft. These were critical to keep the bottom of the shaft from flooding and to provide fresh air for the miners working there. Auxiliary pumps on several platforms were needed to lift water all the way to the top.[2]

On top of the central shaft stood the central shaft building. A two-story wooden structure with a 40 × 60 foot foundation, it completely covered the opening of the central shaft. It contained the steam engines to drive the hydraulic pumps and air ventilator. Another more powerful engine hoisted the iron buckets. This engine and hoist were controlled from an engineer's cabin above the shaft opening. Several hundred cords of wood were stacked inside the building to fuel the steam engines. An array of hand tools and spare parts for the engines were also scattered about. One piece of equipment, referred to as the "gasometer," was housed slightly to the side of the central shaft opening in a small basement room not far from the engineer's cabin. The gasometer was a device for converting naphtha, a crude form of gasoline, into a gas that could be used to illuminate the tunnel. The commissioners had used the gasometer and the piped gas system it supplied for less than two years before deciding it was too dangerous. They abandoned it several months before the new contractors, Dull, Gowan & White, took over work on the central shaft. Unfortunately, no one had explained to the new contractors that the gasometer was considered dangerous and had been mothballed. Consequently, they replenished the gasoline tanks that fed the gasometer and attempted to reactivate it on the afternoon of October 20, 1867, just as James Fitzgerald and his coworkers were sending the first bucket of blasted rock up the central shaft.[3]

A clerk named R. R. Peet, who had worked for the commissioners and was now employed by Dull, Gowan & White, and another man, George Goodwin, were charged with activating the gasometer. Peet carried a lantern into the basement room housing the gasometer when a massive explosion blew both men backward, burning Peet's face. The force of the explosion knocked the engineer out of his cabin, and the metal bucket he had just unloaded tumbled back into the central shaft. Flames quickly spread through the building. The water pumps and ventilator failed immediately. The uppermost platform was loaded with drills, hammers, and chisels. When it gave way, three hundred newly sharpened drills plunged down the shaft like weighty javelins. For the miners at the bottom of the shaft, there was no escape. Since there was no ladder from the work site to the lowest platform, miners had no way of getting out of the shaft once the bucket hoist was rendered inoperable. Those not killed by falling timbers and equipment probably suffocated within the first hour after the explosion. By that time, firefighters from North Adams had arrived and were trying to extinguish the remains of the central shaft building. However, most of its structure and contents had already fallen into the central shaft. At four o'clock, Sunday morning, a brave soul named Thomas Mallory was lowered into the shaft. When he was lifted out, he was unconscious from the fumes at the bottom. After being revived, he reported simply "No hope. No hope." He had seen no bodies floating in the nine feet of water at the bottom. All the miners were gone, probably pinned under the rising water by falling timbers and equipment.[4]

News of the disaster traveled fast. On October 21, the *Boston Post* called the central shaft fire "A Terrible Calamity." The newspaper rated it one of the worst disasters in Massachusetts history. It might have been expected, the *Post* ventured, that after the recent war, the public had "become inured and hardened to the contemplation of [such] horrors." Not so, the newspaper argued. This was a disaster in which "thirteen able-bodied men, full of health and vigor were ushered without a note of warning into eternity." In a final act of respect, the newspaper listed the names of the dead miners: "Patrick Conolley, leaves a wife and seven children; James Burnett; James Fitzgerald, the foreman's brother; Thomas Mulcare; Edwin Haskins; Thomas and Patrick Collins, brothers; Michael Whalan; James Cavanaugh, leaves a wife and two children; John Curron; Thomas Cook, leaves a wife and two children; James McCormick; and Joseph Messier, a Frenchman." Most were young men, from the ages nineteen to thirty. Only two, Patrick Conolley and James Cavanaugh, were over thirty.[5]

In spite of its sympathy for the victims of the central shaft tragedy, the *Boston Post* refused to assign blame. "It is not attributable to the carelessness or ignorance of any human being," the newspaper argued. Not everybody agreed. Benjamin Latrobe believed the casualties at the central shaft could have been avoided "with ordinary care." Francis Bird, James Shute, and other avowed enemies of Alvah Crocker attributed the disaster to the lead commissioner's penny-pinching and careless oversight. There was plenty of blame to go around. Crocker's only remaining engineer, W. P. Granger, left for a new job at first opportunity. The central shaft disaster also ended the relationship of Dull, Gowan & White with the tunnel. Since their contract held them responsible for the equipment lost in the central shaft fire, estimated at approximately $60,000, the firm was forced to abandon the tunnel and soon filed for bankruptcy. However, the tragedy at the central shaft was quickly forgotten. In fact, a week later, a massive celebration in Shelburne Falls marked the official opening of the railroad from Greenfield to that town. Governor Bullock, Alvah Crocker, and nearly a thousand celebrants toasted the future of the tunnel. It would take over a year to recover the bodies of the thirteen miners from the bottom of the central shaft. Their remains were taken to the North Adams Catholic Church and then to the town's Hillside Cemetery. They were buried in a single, long grave there. Since it was left unmarked, the grave cannot be located today.[6]

Following the central shaft disaster and the resignations of Dull, Gowan, & White, Alvah Crocker took back responsibility for work on the tunnel. At the west portal, he could depend on B. N. Farren to push his brick tube further into the mountain. Without a second contractor, Crocker reverted to his strategy of relentless cost-cutting and accelerating progress at the east portal. Inevitably, he incurred sharp criticism from Benjamin Latrobe, who urged him to replace the exhausted and underpowered pumps at the west shaft. Crocker demurred, and work there ceased when the west shaft again became completely flooded. Latrobe complained to Governor Bullock that Crocker was driving the tunnel's east heading "under the specious idea that popular favor will be best propitiated by a mere progress in running feet."[7]

But Crocker's chief critic remained Francis Bird, who delivered a double-barreled blast of pamphlets during 1868. Of the two, *The Modern Minotaur* and

The Last Agony of the Great Bore, the former was the more powerful. In it, Bird compared the Hoosac Tunnel to the "insatiable monster" of Greek mythology, which had demanded an annual tribute of gold and human sacrifice from the citizens of Athens. Every year, Athens had sent ships with black sails to the Minotaur's island cave but never succeeded in quenching the beast's voracious appetite. For Athens, it was both costly and humiliating. It was the same for Massachusetts, Bird insisted. The tunnel would cost the state over $18 million, he estimated, with annual interest becoming the state's largest budget item. Even worse, the tunnel lobby had corrupted the workings of government. This "mischievous element" had leveraged its influence by unscrupulously "forming alliances with every scheme of extravagance and plunder" to ensure the tunnel's continued funding. The modus operandi of this "venal crew" was to identify any state legislator who wanted passage of a particular bill and offer quid-pro-quo deals for tunnel support. It didn't matter to the tunnel lobby what the bill was for—the Horn Pond Railroad, Cape Cod Harbor, or Maverick Bridge—if support for it would result in tunnel funding. Of course, the real monster in all of this was Alvah Crocker. Bird evoked shades of William Shakespeare's *Richard III* (wherein the distraught queen calls out to those brutally murdered by the king) as he listed those individuals who had been sacrificed to Crocker's unbridled ambition and lust for power. What had happened, Bird asked, to Haupt, Brooks, Shute, and Doane? All had passed on leaving Crocker "monarch of all he surveys."[8]

In spite of his obvious drama and hyperbole, Bird was not far off the mark regarding Crocker's lobbying methods. His operation was run out of a set of rooms at 5 Avon Place, not far from the State House. Every Tuesday and Thursday during the legislative session, the tunnel lobby hosted members of the Massachusetts House and Senate for drinks, cigars, and card games well into the night. The operation was directed by E. D. Forster of Cheshire and his first lieutenant, Robert C. Nichols, known respectively as the "General" and "Bob." It was estimated that thirty to forty members of the Senate and three-quarters of the House attended these celebrations regularly. Some of the deals struck with these visitors were significant. For example, the advocates of the Williamsburg & North Adams Railroad got the tunnel lobby's support for a $1 million state loan and the right of towns along the line to subscribe to the company's stock. Similarly, the Boston, Hartford & Erie Railroad received tunnelite backing for a $3 million loan. Smaller deals were also consummated. The town of Amherst was promised votes for a new agricultural college in exchange for approving tunnel funding. Senator William Schouler, who was on the Joint Committee for the Hoosac Tunnel but was ambivalent about the project, gained tunnelite support for a history book he had written and believed deserved to be distributed in every public school and state library. The favor needed to make this happen was small, but Schouler's vote was important. The results of these machinations came to a head in June 1868, when the Massachusetts House voted 115 to 91 to approve $5 million in additional funding for the Hoosac Tunnel. The Senate concurred with a similar margin.[9]

Francis Bird was stunned. In addition to being outmatched by Crocker in the art of logrolling, Bird and his allies were out of step with the emerging politics of the day. They were too rigid and doctrinaire to participate in the "give and take" of politics, having been formed in the high heat of Free-Soilism rather than the cool thought of Whiggery. Their ideology was so tightly harnessed to the antislavery cause that they were labeled "Jacobins" by party moderates. As such, they were no match for the nimble, pragmatic Crocker and his ring of lobbyists. Furthermore, Bird's pamphlets had lost much of their vitality, and the tunnelites countered with a scathing pamphlet of their own. Authored by an anonymous supporter of the Troy & Greenfield Railroad identified only as "Theseus," *The Death of Our Minotaur* mercilessly ridiculed Bird, calling him "the Walpole dyspeptic" who pours out "Pecksniffian lamentations and abuse from [his] never-failing fountain of spleen and remorse." Just to turn the blade in his victim, Theseus quoted a ditty that had circulated in the State House about the stiff and eccentric Bird:

> There once was an old Gander in Walpole,
> Who attempted to sit on a tall-pole:
> Which was very absurd
> In this silly old Bird,
> This funny old Gander from Walpole.

Theseus vilified Bird and his friends at the Western Railroad for their shameless parliamentary maneuvers and political wire-pulling to defend the railroad's cross-state monopoly and thwart competitors. The high cost and slow progress of the tunnel, the pamphlet declared, had as much to do with Bird's "igno-rant interference" as the geological innards of the Hoosac Mountain. Bird had "haunted the State House from the coal hole to the dome," constantly renewing his attack on "the old Boar whose tusks he had felt so often." The real Minotaur roaming the green hills of Massachusetts was Bird, in league with the Western Railroad and its wealthy stockholders in the State House. At a stroke, *The Death of Our Minotaur* neutralized Bird's chief line of attack.[10]

Tunnel funding had always come with strings attached, and the so-called 1868 Allocation Bill was no exception. It required that 20 percent of each pay-ment to tunnel contractors was to be withheld until the tunnel was completed. Interest would be paid on the retained money. The bill also stipulated that the tunnel work could not be broken up between several contractors but had to be performed by just one. It became known as the "single contractor clause." To these burdensome requirements, Governor Bullock added one more: that the contractor selected had to put up a $500,000 security deposit to be returned when the tunnel was completed on time and to the state's satisfaction. Interest would be paid on the deposit. This provision was so egregious that most of the contractors withdrew immediately from the bidding, including Osborne, Gard-ner & Company, the firm favored by Crocker because it was the lowest bidder. Only the Shanly Brothers of Canada remained marginally interested but also threatened to withdraw if the $500,000 security deposit couldn't be renegoti-ated. It appeared to many that Governor Bullock, seemingly supportive of the tunnel in the past, was about to sink the entire project.[11]

Some claimed that Bullock, a native of Worcester, had never been for the tunnel in the first place or had been bought off by the Western Railroad. Others blamed Benjamin Latrobe, who had developed a genuine dislike for Crocker and had drafted the security deposit provision for Bullock. In fact, the gover-nor was in political trouble and trimming his tunnel position. Whereas Bullock had won 77 percent of the vote in his 1866 reelection, he had captured only 58 percent in his third-term victory the following year. His mistake had been deploying Massachusetts' newly established constabulary to enforce the state's outdated liquor law. The law dated back to the anti-Catholic Know Nothing years and, if fully enforced, amounted to virtual prohibition. As Bullock shut down grog shops and bootlegging operations across the state, both Democrats and Liberal Republicans organized against him. Since not only Bullock but other prominent party men like Wendell Phillips and Henry Wilson were also temperance advocates, the Radical Republicans were blamed for the liquor law imbroglio. Many Liberal Republicans joined the Democrats under John Quincy Adams, Jr., hoisting a banner of "Adams and Liberty" and attacking Bullock full force in the 1867 election. This alliance called for more liberal liquor licensing and retrenchment in public spending, especially on the Hoosac Tunnel. Following his weak reelection performance, Bullock backed off enforcement of the state's liquor law and began backtracking on his tunnel support during 1868. This was the problem Crocker and the tunnelites faced as the deadline for tunnel bids approached.[12]

For their part, Bird and the anti-tunnel faction were delighted by these developments. They realized that this might be their last chance to deliver a death blow to the Hoosac Tunnel. After all, less than 40 percent of the tunnel's intended length had been excavated by the end of 1868. In such an unfinished state, the project was extremely vulnerable to being mothballed if no builder could be found to take it on. What is more, if no one picked up the tunnel contract by the end of the year, the new legislature might rescind the $5 million loan to the Troy & Greenfield Railroad in its entirety. With this goal in mind, Bird fought every effort to loosen the onerous provisions of the tunnel contract. He argued that the one-contractor clause had been "written in blood" by the legislature and could never be rescinded. When it looked as though the Shanly brothers might negotiate an alternative to the security deposit, Bird tarred them as "foreigners," usurping business from more deserving Yankee firms. It reminded people of the way he had characterized Herman Haupt as an out-of-state mercenary a few years before. Bird would go on to claim that the state was grossly overpaying the Shanlys compared to what Dull, Gowan & White had received for their work. The governor and his council, Bird argued, were being duped by Crocker's blatant falsehoods and "pettifogging acumen." Another strident voice in the anti-tunnel faction, William Robinson, accused Crocker and his cronies of plying legislators with liquor at their "assignation house" on Avon Place. He commissioned a special legislative investigation into Crocker's lobbying methods. It turned up nothing improper, since most of the investigators had visited Avon Place numerous times. There was considerable hypocrisy in all of this, and the public recognized it. Both Bird and anti-tunnelite Richard Henry Dana, Jr., who had vehemently opposed funding for the Troy

& Greenfield Railroad, had no compunctions voting for a $3 million loan to the Boston, Hartford & Erie Railroad. It was revealed that Bird's paper mill shipped its products over that road and Dana's business partners were heavily invested in it. During all of this, Walter Shanly persisted. He visited North Adams for a closer look at the tunnel and enlisted the support of the town's business community. In seemingly endless negotiations with Governor Bullock and his Council, Shanly agreed to do $500,000 worth of "free work" up-front instead of putting down a security deposit. Benjamin Latrobe strongly objected to the idea and resigned in protest as the governor's consulting engineer. Shanly later remembered Latrobe as "the hardest man I ever had to deal with." Following Latrobe's departure, negotiations moved ahead more smoothly and were brought to a conclusion by Shanly's lawyer, Elias Hasket Derby. By this time, Derby was the foremost railroad lawyer in the state and had gotten over his poor treatment from Herman Haupt. A final contract was signed between the State of Massachusetts and Shanly Brothers on December 24, 1868. The Shanlys agreed to complete the Hoosac Tunnel for the sum of $4,598,268 with a deadline of March 1874. The future governor could extend this by six months.[13]

In spite of the roadblocks he had thrown in the way of the Shanly contract, Governor Bullock turned positive on the new contractors and the tunnel in his January 1869 valedictory address. The state had imbedded "satisfactory guarantees" in the contract to insure timely and efficient completion of the tunnel, he explained. The state's investment in the project would result in "far-reaching connections which ally the manufacturing and commerce of the East with the granaries of the West." New York and Pennsylvania had already made these kinds of investments and thereby secured their economic destinies. "Massachusetts could not afford," the governor warned, "to cherish a policy less broad than that which has conducted them to prosperity and greatness." The state now seemed committed to the tunnel as never before. Governor Bullock's words could have been scripted by Alvah Crocker and Elias Hasket Derby eighteen years earlier.[14]

The Hoosac Tunnel had been tossed about on the roiling sea of politics for years. Sometimes politics broke for the tunnel and sometimes against it. The election of 1868 favored the tunnelites. Whereas the dominance of the state's Radical Republicans had eroded over the previous two years—with local issues like liquor law enforcement and public spending on railroads drawing attention away from national concerns about southern reconstruction and black enfranchisement—the elections of Ulysses S. Grant to the presidency and William Claflin to the governorship of Massachusetts sharply reversed this trend. After almost four years of Andrew Johnson's presidency and failed attempts to remove him from office, the Radical Republicans were eager to have one of their own as the nation's chief executive. What is more, the national Democratic ticket, consisting of Horatio Seymour, the wartime governor of New York State, and Francis Blair, Jr., a corrupt and reactionary Missourian, was poorly constituted. Seymour was a hard-money man with eastern business connections, who worried soft-money midwesterners and southerners. His running mate, Blair, was even worse. He spewed antiblack rhetoric with so little restraint that he offended many moderate Democrats and independent voters. During the election, broad-scale violence and black voter intimidation in the South horrified many northerners. Grant won 53 percent of the popular vote and swept the Electoral College 214 to 80. The Massachusetts Republican Party avoided all state and local issues during the election and allowed Claflin to ride into the governorship on Grant's national coattails. Claflin had served as lieutenant governor for three years under Bullock and understood how gingerly he needed to handle issues like liquor legislation and railroad funding once in office. As a native of Milford, Massachusetts, south of Boston, he had no predispositions in favor of any railroad. He was a leading shoe manufacturer in the state and chairman of the National Republican Committee. As a member of the Bird Club, he had its support for the governorship. Already a wealthy man, he was a "disinterested" politician who sought factional peace within his party. Claflin would hold Shanly Brothers strictly accountable for their work on the tunnel and veto funding for the Boston, Hartford & Erie Railroad. Though his mandate from voters would erode over time, Claflin's smart, balanced handling of state railroad funding and other issues would see him through three terms in the governor's office.[15]

Though more nuanced toward tunnel ideology than his predecessor, Governor-Elect Claflin touched on several fundamental points concerning it in his inaugural address. He picked up the ancient lament regarding the poor natural endowments of the Bay State. "Nature has done so little for us in comparison to the grain growing region of the West," he complained. The new governor's reference to western grain hinted at a tantalizing solution to the problem. Avoiding specifics, Claflin stressed the importance of transportation to the state's future. "Few things are of greater importance to the community," he continued, "or a surer test of civilization than good roads." Here, Claflin was tapping into the long-standing belief that good roads removed the debilitating effects of isolation and connected scattered communities to progress and culture. For the governor's more contemporary audience, Claflin was understood to be talking about *good railroads*.[16]

Walter Shanly and his younger brother Francis had emigrated from Dublin, Ireland, to Canada in 1838 at ages nineteen and sixteen, respectively. Their father believed strongly in education and saw to it that his sons translated their share of Latin verse before leaving Dublin for their new life. They worked on their father's farm west of London, Ontario, long enough to know they wanted to do something else. Both became civil engineers working for a succession of canal and railroad companies. They were essentially self-taught, absorbing available engineering textbooks and learning quickly on the job. Walter rose to become chief engineer on the Toronto & Guelph Railroad and then general manager of the Grand Trunk Railroad. He was elected to Parliament for several terms and invested in a bank as well as a starch company. His brother Francis, the more fun-loving of the two and a family man, developed a reputation for assiduous bookkeeping and complex project management. Walter remained a bachelor and loved solving difficult engineering problems working in the field. One was an inside man and the other an outside man. When the opportunity to bid on the Hoosac Tunnel arrived, the two brothers had not worked together for fifteen years. They quickly cobbled together the firm of Shanly Brothers on paper, trusting in each other's complementary skills to land the job.[17]

Walter Shanly became the public face of the firm. In photographs of him, his aquiline features and riveting gaze stand out. Though he was usually reserved and at times aloft in manner, people still took to him and were impressed by him. Because of his experience in the field, he was especially adept at communicating with the working men on various project sites. He at once exuded authority and empathy. A reporter for the *Adams Transcript* described Walter Shanly as "striking and commanding." He exhibited "an energy and determination which will doubtlessly enable him to solve the problem of the Hoosac Tunnel," the reporter wagered. Shanly's confidence in the tunnel was catching. In a speech before the Boston Board of Trade in 1870, with Governor

Claflin in attendance, Shanly predicted that "in three years from now we can run a train, almost on a dead level, from here to the Hudson." He received a standing ovation.[18]

As had frequently been the case in the stop-and-go history of the tunnel, recommencing work under the Shanlys seemed to take too long. The firm's approach was thorough and methodical. Everything had to be right before actual work could begin. Nor could construction start until the brothers had secured a $600,000 loan from their Canadian bankers. With that money in hand, the Shanlys could negotiate the purchase of the state's equipment at the tunnel site. Much of that equipment was dilapidated after years of use and Crocker's penny-pinching. It was necessary to upgrade engines and pumps in several locations and rebuild the miners' living quarters to better standards. Fortunately, the central shaft had been bailed out and the brick tube at the west end completed by early 1869. Still, the central shaft building had to be rebuilt with an eye to improved safety and the platforms inside the shaft completely reconstructed. The Shanlys installed a fireproof floor with self-closing iron hatches in the central shaft building. More than 150,000 board feet of lumber were used to construct sixty platforms inside the shaft. A shortage of carpenters delayed this work. Organizing the work force also took time. The Shanlys acted purely as project managers and subcontracted all work on the tunnel. The three main subcontractors were Dawe & Dobson, Doisgenoit, and Callihan. Contracts with these firms needed to be negotiated and their laborers brought to the tunnel. Work on the east portal began at the end of March 1869 and on the west portal a few weeks later. Work at the central shaft was delayed until late May.[19]

In spite of these delays, the Shanlys were fully engaged at the tunnel by summer 1869 and gathering momentum. By that time, over nine hundred men were working on the tunnel in three eight-hour shifts. Governor Claflin visited the tunnel during July 1869 and came away impressed. By September, the Shanlys had hard data to validate the governor's optimism. Miners at the east heading were penetrating the mountain at the unprecedented rate of 160 feet per month. The Shanlys fully embraced the Burleigh drills, Mowbray's tri-nitroglycerin, and the Browne fuses, all advantages that had been denied to Herman Haupt. The Shanlys mounted four Burleigh drills on each drill carriage and placed two carriages side by side at each heading. Miners drilled blast holes four to five feet deep in a sideways-pointing V-formation, a drilling pattern still used today. This allowed the nitroglycerin to tear away the widest possible hole in the heading. The brothers also replaced mules at the east portal with a small steam locomotive to speed removal of spoils from the tunnel. They shrewdly christened the little locomotive "Gov. Claflin." At the west end of the tunnel, where it was necessary to hoist spoils up the west shaft in buckets, miners achieved an average monthly penetration of 120 feet, double the previous record. In the central shaft, where hand drilling was still used because of its verticality, workers were going thirty-three feet deeper every month. At that rate, the central shaft would reach tunnel grade in a year.[20]

Optimism ran high in North Adams, Greenfield, and even Boston. However, cautious observers recalled the experience of Dull, Gowan & White and how abruptly the fire at the central shaft had ended their tenure at the tunnel. This time, it was not fire but water. On Sunday afternoon, October 3, 1869, a torrential rain began to fall in northwestern Massachusetts and continued for over a day. An estimated seven inches of rain caused rivers to burst their banks and inundate the main streets of area towns. In Shelburne Falls, the Deerfield River rose six feet and tore away the town's bridge. The "Great Flood of '69," as it was later called, was an extraordinary weather event and a disaster for the Hoosac Tunnel. On Monday morning, October 4, near the tunnel's west end, a stream that had been diverted away from the west portal overflowed its banks and headed in the direction of the tunnel. For a short time, water and debris flowed into the cavity left by the old Haupt entrance but, once it was filled, poured into the brick tube of the west portal. About seventy-five men were working there and fled for their lives through the adit connecting the west end of the tunnel and the west shaft. Fortunately, it was daylight and James Hocking, one of the tunnel's subcontractors, saw what was happening and rode up the mountain to the west shaft. He warned the crew manning the hoist there to prepare to extract the miners below. He would be lauded by the local press as the day's hero. With the flood water rising in the west shaft, all but one of the miners were lifted to safety. The only casualty, Richard Barryman, became confused about the route to safety and drowned coming through the adit. It was his first day on the job. The force of the water at the west end washed out the backfilled earth around the brick tube and damaged the timber galleries behind it. Fortunately, B. N. Farren's brick tube held together. Still, massive amounts of debris clogged the west portal and made it impassable. At the east end of the tunnel, there was less threat to life but considerable damage to the railroad tracks leading to the tunnel. The bridge over the Green River was completely destroyed and a considerable amount of roadbed washed away along the Deerfield River. The trackage from Greenfield to the east portal would take nine months to repair. The lower section of the Deerfield Dam was undercut but the dam itself remained intact.[21]

Unlike the catastrophe that had befallen their predecessors, the Flood of '69 turned out to be a test of the Shanly Brothers' resilience and not a terminal event. Of course, the two events were different. In the case of Dull, Gowan & White, it had been the loss of equipment that was so financially devastating. Then as now, engines and motors were the most precious assets on a construction site. The Shanlys lost far less equipment than their predecessors had. For the Shanlys, recovery entailed a major clean-up at the west portal and some repairs to the tunnel's infrastructure there. The locomotive "Gov. Claflin" was hauled over the mountain by a team of mules to help clean out the flood debris from the western end of the tunnel. The brothers were not responsible for the wrecked railroad track to the east portal. That job fell to the state. The Shanlys were also better capitalized, allowing them to absorb any equipment losses and begin repairing damage to the tunnel quickly. Importantly, there was minimal loss of life from the flood and no blame for it was assigned to the Shanlys. Contrary to expectations, the Shanly brothers did not cease working on the tunnel after the flood. They appealed to the state for financial relief but pushed ahead while waiting for reimbursement. They were widely praised for this. Still, repairing damage to the west portal and removing debris from it were expenses the brothers had to cover themselves. So, too, was the extra cost of transporting supplies for work at the east portal over the Hoosac Mountain while the railroad from Greenfield remained out of service. In the end, the Shanly brothers not only survived but carried on. In doing so, they signaled to the public that, unlike the hand-to-mouth contractors before them, the Shanlys were engaged in the tunnel for the long haul and likely to complete it. The Shanlys' quick recovery and perseverance after the flood turned public opinion strongly in their favor.[22]

So, too, the decades-long struggle to secure adequate and sustained funding for the Hoosac Tunnel had come to an end. The Shanlys had their contract firmly in hand, and it seemed sufficient to ensure the project's completion. After the contract was signed, Francis Bird retired from the legislative battlefield over tunnel funding and discontinued his stream of anti-tunnel pamphlets. Perhaps, he had decided to focus his attention on a rejuvenated Western Railroad. In 1868, that line had finally merged with the separately managed Boston & Worcester to create a unified Boston & Albany Railroad. The merged railroad increased its capitalization by $2 million to fund double tracking, new rolling stock, and the easing of curvatures through the Berkshire Mountains. An improved connection to Boston's wharves and a grain elevator there were planned. Completed in 1866, a bridge over the Hudson River at Albany favored the line. Left unaddressed were two important questions that would also affect the Hoosac Tunnel route once it was finished: How would Boston connect with Europe after the Cunard line began bypassing Boston in the late 1860s in favor of New York City? With that development, Boston lost its only ocean-going connection to Liverpool, England, and European markets. Also, how would Boston control the western shipment of grain and other goods across New York State on railroads over which it exercised no authority? These railroads were owned by New York corporations and naturally favored that city for the export of goods they carried. These larger questions, like so many having to do with the Bay State's hopes of ending its commercial isolation, hung menacingly over the day-to-day preoccupations of the two emerging cross-state railroads.[23]

Francis Bird was also consumed with politics. He was desperately trying to hold the Massachusetts Radical Republicans together as the Civil War faded into memory. The Democrat-leaning *Boston Post* accused Bird and his radical allies of "raking over [their] negro embers to find warmth in them." In doing so, the Radical Republicans continued to neglect local and state issues such as liquor legislation, workingmen's hours, and retrenchment in public spending. As a consequence, the radical consensus gave way to intraparty factionalism and the rise of splinter parties. Governor Claflin's margin of victory declined in both 1869 and 1870. Furthermore, President Grant's feud with U.S. senator Charles Sumner—the iconic leader of the Massachusetts radicals—would drive Bird, Nathaniel Banks, William Robinson, and other Bay State radicals into previously undreamed-of alliances with Liberal Republicans and Democrats. By the early 1870s, Massachusetts politics would be scrambled beyond recognition and voter turnout depressed by public frustration and cynicism.[24]

One positive development of the period was the establishment of the Massachusetts Railroad Commission in 1869. It had taken four years of legislative effort by railroad reformers to make it a reality. Charles Francis Adams, Jr., great grandson and grandson of presidents, Union Army veteran, Harvard graduate, and journalist, assumed its leadership. With his brother Henry, he had written *Chapters of Erie*, a "muckraking" expose about Cornelius Vanderbilt, Jay Gould, and other speculators in their struggle for the Erie Railroad. In the book, Adams described railroads as "an enormous, incalculable force practically let loose suddenly upon mankind." Under Adams's leadership, the new commission would make recommendations on railroad safety, financing, and freight rates. It would be partly responsible for lowering freight costs in Massachusetts during the 1870s. Adams and his commission would also play an important role in deciding what to do with the Hoosac Tunnel once it was completed.[25]

Neither the Bay State's morphing political landscape nor its long-term commercial vision concerned Walter and Francis Shanly. They were focused on driving progress at the Hoosac Tunnel and getting paid for it on a reasonable schedule. Regarding the former objective, 1870 proved a banner year. The tunnel's eastern heading was advanced a record 1,514 feet and the western heading a record 1,203 feet. In total, they had increased the length of the tunnel by almost half a mile and nearly reached its halfway mark. Moreover, the central shaft reached tunnel grade in mid-August. At 1,028 feet, it was the deepest hole man had ever dug. Regarding the Shanlys' getting paid for their work, things went less smoothly. The problem was James Laurie, who had taken over the job of consulting engineer after Benjamin Latrobe's departure. Laurie had served as one of the civil engineers hired by the original tunnel commissioners back in 1863 and had contributed to their report to Governor Andrew. For whatever reason, Laurie became highly critical of the Shanlys and a major thorn in their side. First, he attempted to foist an upside-down payment schedule on the brothers which would remunerate them, not for work completed, but rather for the total contract price less the value of work left to be done. Such a schedule did not take into account unanticipated expenses such as the clean-up after the previous year's flood or the cost of new equipment as the tunnel progressed. When he failed to impose this payment system on the brothers, Laurie found fault with their grading and surveying methods. The tunnel's superintending engineer, Benjamin Frost, believed Laurie had lost his mind. Given

Laurie's obsessive and inexplicable behavior, perhaps he had. When Laurie and Frost became deadlocked in a dispute over a surveying question, Governor Claflin lost patience and fired Laurie. His replacement was Edward Philbrick, a Boston civil engineer with a good reputation, who sided with Frost and verified the accuracy of the Shanlys' survey. For the moment, Philbrick cleared the air regarding Laurie's criticism of the Shanlys and allowed his predecessor's proposed payment scheme to be taken off the table. In the longer term, however, Philbrick would show poor judgement in his oversight of the tunnel and nearly cause a catastrophe with his reckless exercise of authority.[26]

Laurie's criticism of the Shanlys' surveying methods was ironic, given that the brothers were about to carry out the most impressive surveying achievement in the tunnel's history. The prime mover in this was Carl O. Wederkinch, who had taken over as superintendent of the central shaft following the death of the diligent and accomplished Cornelius Redding. Redding and two of his colleagues were killed when the rope on the central shaft bucket broke, sending them to the bottom of the shaft. They either died on impact or drowned in the standing water there. Wederkinch himself was nearly killed by falling debris soon after assuming Redding's duties. Wederkinch had graduated from the University of Copenhagen and emigrated from Denmark to New York not knowing a word of English. He worked as a common laborer while acquiring the language, arriving at the Hoosac Tunnel during 1870. Fortunately, Walter Shanly recognized the innovative genius behind Wederkinch's drab mining outfit and halting speech. It would fall to Wederkinch to figure out how to survey the tunnel line east and west from the bottom of the central shaft, more than a thousand feet below ground.[27]

Thomas Doane had projected the tunnel line from adjacent mountain peaks across the east and west portals and placed the central shaft as close to halfway between them as topography would allow. He had also calculated the proper depth of the central shaft, in order to create a gentle slope toward both portals. That configuration was meant to encourage smoke from trains in the tunnel to vent up the central shaft. It would also allow spoils to be carted to the portals more easily and encouraged drainage out the portals. The challenge now was to project the tunnel's azimuth so accurately from the bottom of the central shaft that the two headings emanating from it would eventually meet up with the headings coming from the east and west portals. Given the distances involved, the slightest inaccuracy would result in miners blindly digging past each other deep inside the mountain.[28]

Wederkinch first had to verify the tunnel's precise centerline across the top of the central shaft. He built two stone piers, one twenty-five feet east and the other the same distance west of the central shaft's elliptical opening. On top of each pier he mounted a device known as a "vernier," each with an adjustable, calibrated vertical slit through which he could make precise sightings and record them in his notebook. Through these vernier slits, Wederkinch focused on Thomas Doane's mountaintop surveying monuments on Row's Head to the east and on Notch Mountain to the west. Miners removed siding from the central shaft building to allow him a clear visual path. Wederkinch found that atmospheric conditions caused minute variations in his sightings, for which he carefully adjusted his vernier slits and recorded the corresponding calibrations. After numerous sightings, he averaged his calibrations to compute what he believed was the true line of the tunnel. Wederkinch then stretched two parallel steel wires through the slits of his verniers, each wire touching a side of the slit set to his average sighting. These two wires were 1/20th of an inch apart and stretched taut from one vernier slit to the other. Next, he suspended two copper wires from a framework above the taut steel wires down between them, not allowing them to touch the steel wires. Each vertical copper wire was suspended eleven and a half feet from the center of the central shaft opening, or twenty-three feet apart. The two copper wires reached down more than a thousand feet to the bottom of the central shaft. Because of the earth's movement at that depth, drafts in the shaft, temperature changes, and spider infestation—the two copper wires oscillated wildly. Wederkinch enclosed them in two eight-inch-square wooden boxes running the length of the shaft and set their plumb bobs at the bottom in buckets of light oil. He reduced the oscillation to 1/100th of an inch. At the bottom of the shaft, Walter Shanly set up a platform on which Wederkinch mounted a second set of verniers and sighted the subterranean tunnel line through the two vertical copper wires. Again, he made dozens of sightings and averaged them before selecting what he thought was the exact east-west line of the tunnel below ground. In this way, Wederkinch transposed the tunnel's surveyed centerline from the surface above to the bottom of the central shaft. Wederkinch later patented his special surveying equipment, without which he could not have determined the tunnel's subterranean line. Though his notebooks have been lost, he is believed to have made hundreds of sightings. In the end, his efforts paid off. When the central shaft's east heading met up with the heading coming from the east portal, the two tunnel lines diverged by only 5/16th of an inch. The west heading of the central shaft was only 9/16th of an inch off the line from the west portal. Once word of his accomplishments spread, the American mining industry adopted Wederkinch's methodology and proprietary surveying equipment.[29]

The unique challenges of the Hoosac Tunnel made it a hot house for technical innovation. Whether in pneumatic drilling, use of tri-nitroglycerin, electrically ignited fuses, or subterranean surveying—the special problems the builders of the tunnel faced and overcame spawned new devices and methods that would prove as historically significant as the tunnel itself. As they were refined and commercialized, many of these technical advances gave rise to whole new industries. After the death of Cornelius Redding and his fellow miners, for example, the Shanlys decided to replace the dangerous bucket hoist in the central shaft with a safer system. A more efficient system was also needed to remove the increased amount of spoils from the two new headings opened up at the bottom of the shaft. The new system consisted of a capacious elevator cage riding in fixed guiderails attached to the walls of the shaft. The Shanlys also installed a more powerful hoist engine in the central shaft building. The new elevator could lift three cubic yards of spoils or a dozen miners to the surface. Most important, the elevator's safety device was mechanically simple but lifesaving for the miners. A horizontal, bow-shaped elliptical spring on top of the elevator was attached to the lift cable and compressed when the elevator was raised or lowered. When the tension on the lift cable was relaxed, the spring expanded sideways to grip ratchets inside the elevator's guiderails. In that position, the elevator was locked in place until tension was restored to the lift cable. In other words, the elevator could not fall if its cables broke. The inventor of the safety device was Elisha Otis, and its first commercial application was in the Hoosac Tunnel. While it is unclear whether Otis designed the entire elevator, his safety device would revolutionize vertical conveyance in mining operations and, later, urban architecture.[30]

In spite of the impressive engineering accomplishments in the central shaft, Walter Shanly was never convinced of its necessity to the overall tunnel plan. He questioned both its ventilation capability and the need for two new headings to finish the tunnel. Furthermore, the central shaft was the most dangerous work site at the tunnel and always problematic for its builders. Nonetheless, completing the central shaft was a requirement in the Shanly contract. For every challenge overcome in the central shaft, a new one seemed to present itself. Flooding in the central shaft had always been a problem. Lifting water a thousand feet up the shaft had to be done in stages and required a series of pumps and cisterns on several of the shaft's interior platforms. Once brought to the surface, pumped-out water had to be channeled a thousand feet from the shaft opening to a swamp near the Cold River. During March 1871, the problem of water seepage in the shaft grew worse. When the shaft's west heading reached 180 feet, the composition of the rock changed from mica schist to a dark granite-like material with cracks in it. It was a bad omen. Water began spouting from these cracks and soon exceeded three thousand gallons per hour. The increased flow of water overwhelmed the central shaft's pumping system. Miners persevered in knee-deep, cold water. However, rising water made drilling and igniting explosives difficult. Miners feared that the next explosion might open up an even larger pocket of water which would flood the tunnel and be life-threatening. The Shanlys finally halted work at the shaft's west heading and began searching for more powerful pumps. It would take six months to purchase and install them.[31]

Spending more money on the central shaft, in which he didn't believe in the first place, galled Walter Shanly. Two large Cornish pumps and a rebuilt foundation to support their extra weight cost the Shanlys almost a quarter of a million dollars. Adding to the brothers' frustration, further excavation of the shaft's west heading released fresh reserves of underground water. The new pumps soon proved inadequate to deal with these inundations. By the end of May 1872, the flow of water had increased to eight thousand gallons per hour. Once again, work had to be stopped. Disregarding the obvious risks involved, Consulting Engineer Philbrick objected to the stoppage and ordered that work be recommenced. Over the summer of 1872, the Shanlys argued before the joint legislative committee on the tunnel and tried to convince the new governor, William Washburn, that probing the seemingly abundant quantity of underground water behind the shaft's west heading could jeopardize the entire tunnel project. The Shanlys recommended in the strongest language that the shaft's east heading be driven with an augmented labor force to meet up with the heading approaching from the east portal. Once the two headings were joined, they argued, the flow of water from the shaft's west heading could easily be pumped over the low crest of the tunnel floor at the bottom of the central shaft and drained out the east portal into the Deerfield River. In spite of the logic of this argument, the joint legislative committee, now firmly under the influence of Philbrick, again ordered the Shanlys to push ahead with the shaft's west heading. If recent events were a harbinger of what might happen next, then the Hoosac Tunnel was on the verge of catastrophe.[32]

At this point, Walter Shanly met privately with Governor Washburn. Shanly told him that he and his brother had decided to ignore Philbrick and the joint committee and cease working on the shaft's west heading. If the governor intervened and forced work to be restarted, Shanly warned, the tunnel would be inundated and might never recover. If that happened, the responsibility would rest on the governor's shoulders. Without a detailed account of the meeting, it is impossible to reconstruct the tone of what was said. However, we know Walter Shanly was a strong personality and often direct to a fault. In photographs of him, Shanly's gaze is intense and penetrating. He found himself in an untenable position because of Philbrick's obstinacy and clearly placed the governor under pressure. Governor Washburn was a resident of Greenfield, Massachusetts, and wanted to see the tunnel completed. In the end, he agreed to remain silent while the Shanlys went ahead with their plan.[33]

Had it not been for Walter Shanly's courage, the story of the Hoosac Tunnel would probably have come to a sad end sometime in 1872. In November of that year, what became known as the Great Fire of Boston would destroy sixty-five acres in the city's downtown and cause $73 million in property damage. Less than a year later, the Panic of 1873 would bring on the worst economic depression in the nation's history. Massachusetts would not be spared. Under such circumstances, another disaster at the tunnel would probably have been too much for the state legislature and voting public to stomach. The Hoosac Tunnel, whose survival had always been tenuous, would probably have been given up for good. It was a disaster narrowly averted.[34]

Following this covert agreement between the governor and his contractor, it seemed clear that the Great Bore would be completed. Walter Shanly's sprint to the finish line would lend an exhilarating tone to the tunnel's final chapter. And, yet, the accelerated progress on the tunnel had tragic consequences for the workers who drove it. Whereas there had been sixty-nine deaths and serious injuries at the tunnel during the fourteen years from 1855 to 1869, that number rose to 123 during the six years of the Shanlys' tenure from 1869 through 1874. Since the tunnel was less than 40 percent excavated when the Shanlys took on the job, the toll on workers was roughly proportional to linear feet of tunnel dug during these two periods. Still, the causes of deaths and serious injuries—frequently from falling rock and mishaps with tri-nitroglycerin—indicate that the work had become more dangerous. The increased pace of work, the instability of the ceiling in the western part of the tunnel, and the lack of experience with new technology, especially tri-nitroglycerin, all contributed to the large number of accidents.[35]

At some point during this period, a poem entitled "Only A Tunneller" became popular and was rendered into song. "Only a tunneller passing your door, in a solemn hearse to-day," it begins. The subject of this sad tale is a fictional miner named Tom Lynch, who gives up his life drawing out an unexploded nitroglycerin charge, because his brother Joe is about to arrive on the next shift and might have to do it himself. Joe is a family man with a dozen children. "A low, deep rumble 'tis all that's known of the launch of the tunneller's bark," the song concludes. As the tunnel workers and residents of nearby communities grew closer, the frequency of accidents during the Shanly years left its mark on these towns. After decades living side by side, villagers had become accustomed to seeing the miners and had made acquaintance with some of them. The men who worked at the tunnel were no longer as foreign as they had once seemed. Each earth-shaking nitroglycerin explosion or call for a doctor to amputate a miner's limb elicited sympathy for the hapless victim and his soon-to-be marooned family. The funeral of James Mulaney, who was a Civil War veteran and had labored at the tunnel for nine years, was representative of this developing relationship. Mulaney had died in a nitroglycerin explosion while trying to rescue a coworker. Mulaney was widely popular and a large crowd of townspeople and miners accompanied his remains from St. Francis Catholic Church in North Adams to Hillside Cemetery. A military band supplied the dirge to which the procession kept step. The *Hoosac Valley News* mourned the loss of so many good men at the tunnel and wrote what many felt—"Thank God the work is nearly finished."[36]

7

BREAKTHROUGH AND PANIC

AT FIRST, IT WAS barely audible. But when they heard it, they knew what it was. The excitement among the miners was palpable. There were few moments of silence to listen for those tunneling toward them from the opposite direction. The deafening racket of their own Burleighs, the hiss of compressed air, the constant drip of water, as well as the shouts and curses of the miners themselves made it difficult to hear anything. From time to time, they would stop work and listen. There it was again—the dull thump of an explosion not far ahead and the hammering of drills in the rock wall separating them. By now, everybody in North Adams and many around the country knew the final breakthrough at the Hoosac Tunnel was close. Reporters gathered from as far away as Chicago to record the event, now anticipated for November 27, 1873, Thanksgiving Day.[1]

There were other news stories in the weeks before the final breakthrough and not all of them good. In the middle of September, a New York bank owned by railroad speculator Jay Cooke suspended payments. A financial panic ensued and soon spread throughout the nation, cutting off credit to both large and small businesses. A major cause of the Panic of 1873 was the collapse of railroad stocks, where overspeculation and financial manipulation had run rampant. Both the tunnel breakthrough and the panic were railroad stories, one triumphal and the other sordid.[2]

At the time of the fateful meeting between Walter Shanly and Governor Washburn—on whether or not to recommence work on the central shaft's flooded west heading—these events were eighteen months in the future. In the meantime, 1872 would be a year of remarkable progress at the tunnel. With almost seven hundred miners working below ground and another one hundred or so supporting them at the surface, the Shanly brothers excavated a record 4,456 feet during the year. This was well ahead of the 3,553 feet achieved the year before. Interestingly, the Shanlys took different approaches excavating the portion of the tunnel east of the central shaft versus the portion west of it. In the eastern portion, miners employed a "bottom header" approach, with an advance drilling team working at tunnel grade and carving out the lower eight to ten feet of the tunnel's intended twenty-foot height. A second drilling team followed them on an elevated, rolling bench and "sloped out" the tunnel to its full height. In the western portion of the tunnel, miners did just the opposite. The first team advanced along the roof of the tunnel in a "top header" approach, allowing the team behind them to excavate down to the floor of the tunnel. They did this for safety reasons. Because the western end of the tunnel was notoriously unstable, the lead miners had to make sure the ceiling of the tunnel would not collapse on them before moving on. Where the ceiling was found to be unstable, miners installed wooden trusses to shore it up. This was tedious and dangerous work. Once the drilling teams passed by and the dimensions of the tunnel were trimmed out, a crew of masons constructed brick arching in these unstable sections. Even though B. N. Farren had already installed 931 feet of arching in the western portion of the tunnel, the Shanlys were forced to add another 1,043 feet to protect it from cave-ins of loose rock, incursions of porridge stone, and other hazards. This was four times the amount of arching called for in their contract.[3]

The need for additional arching inside the west portal and the work stoppage at the west heading of the central shaft meant that penetration of the mountain progressed more slowly from that direction. As a consequence, the Shanlys focused their main energy on the eastern portion of the tunnel, where excavation was less problematic and could be driven more rapidly. Every contractor before them had done the same. Fortunately, the Deerfield River remained ice free and ran high during 1872. This abundance of flow spun the turbines just below the Deerfield Dam and powered the air compressors that fed the Burleigh drills inside the east portal. This obviated the need for expensive firewood to stoke the steam engines used when the Deerfield suffered from low water. Importantly, the Shanlys were highly motivated to reach the central shaft from the east. Once that happened, the water inundating the west heading of the central shaft could be drained down the eastern portion of the tunnel and into the Deerfield River. Until then, the Shanlys had to run the pumps on top of the central shaft at full capacity to keep the central shaft from filling with water and endangering miners working its east heading. These pumps consumed six hundred cords of wood per month, a cost the Shanlys could scarcely afford. One legislator asked Walter Shanly why he didn't simply let the central shaft fill up with water and then tap it when miners reached it from the east. Shanly patiently explained that no miner in his right mind would attempt to break into a central shaft containing several hundred feet of water. Ignoring such naïve suggestions, the Shanlys attacked both headings in the tunnel's eastern section with their Burleigh drills. Charles Burleigh, the man Alvah Crocker had characterized as "a monument to genius," had continued to perfect his drills. They now proved their worth. The Shanlys anticipated connecting the east portal and the central shaft—in what would be called the first breakthrough—six weeks ahead of schedule on December 12, 1872.[4]

That event had two consequences: first, Walter Shanly (his brother had left the project some months before) shifted the majority of his resources to the western portion of the tunnel; and, second, serious legislative debates began about what to do with the tunnel once it was completed. As expected, the breakthrough between the central shaft and the east portal allowed the cascade of water from the shaft's west heading—now one thousand gallons per minute—to drain harmlessly downhill to the east portal and into the Deerfield River. Shanly dismantled the large pumps at the top of the central shaft and transferred much of the equipment there to the west portal and west shaft. (Smaller pumps could be relied upon to move water over the slight crest at the bottom of the central shaft so that gravity would carry it down to the east portal.) Since the west portal and west shaft had finally been connected in July 1872 and most of the approximately 2,500 feet between them properly arched, spoils could now be carted out the west portal instead of having to be lifted up the west shaft. Miners immediately began breaking records. The monthly average of excavation from the west heading of the central shaft was 154 feet, three times what the commissioners had seen achieved during the 1860s. After so many problems in its western portion, miners were now engaged in a mad dash there to complete the tunnel. During September 1873, miners working both headings broke all previous monthly records by excavating 316 feet. By that time, less than a thousand feet separated miners digging toward each other inside the mountain.[5]

The second consequence of the first breakthrough was the commencement of hearings regarding the future management of the Hoosac Tunnel and the railroads leading to and from it. This so-called Great Debate of 1873 took place at the State House during February and March. It was attended by a number of distinguished railroad experts, politicians from constituencies through which the Hoosac Tunnel route would run, and representatives of the railroad companies that comprised the route. Beginning with Governor Washburn, there was broad consensus that some form of consolidation was needed in order to join the various railroads along the Hoosac route into one through-line that would eventually use the tunnel. After all, Massachusetts had forced the merger of the Boston & Worcester and Western Railroad a few years earlier. The state had achieved this by threatening to build parallel tracks (known as "nuisance lines") along key portions of their roads if the companies refused to cooperate. However, the merger of the railroads along the tunnel line would prove more complex. To begin with, there were six separate railroad companies involved: the Fitchburg Railroad, running fifty miles from Boston to Fitchburg; the Vermont & Massachusetts, spanning the fifty-six miles from Fitchburg to Greenfield; the state-owned Troy & Greenfield, its tunnel and tracks leading to and from it, totaling forty-four miles when completed; the tiny Southern Vermont, its six miles cutting across the southwestern corner of Vermont to the New York State border; and, finally, the Troy & Boston, covering the thirty-five miles to Troy, New York, on the Hudson River. The hearing took an unexpected turn when the Fitchburg Railroad balked at the idea of consolidating with the other railroads on the Hoosac route. Attendees at the hearings were surprised by this, given how instrumental the Fitchburg Railroad's founder, Alvah Crocker, had been in the tunnel project. Crocker did not attend the hearings, and it is not known what his views were. Crocker had recently lost his wife and been elected to the U.S. House of Representatives. The spokesman for the Fitchburg Railroad was none other than Elias Hasket Derby, who argued that the proposed merger of the railroad with the weaker lines along the Hoosac route would jeopardize the Fitchburg's 8 percent dividend to its stockholders. The only way the railroad would consider consolidation would be if the state allowed the Fitchburg to merge with the profitable Boston & Lowell Railroad at the same time. The latter railroad ran north from the Fitchburg line to Port-land, Maine. Possibly, the Fitchburg's resistance to consolidation was a ploy to accomplish this merger. Or, possibly, the railroad was worried about the costs that lay ahead to bring the Hoosac Tunnel into full operation. Why not let the State of Massachusetts pay those costs and then purchase the finished tunnel?[6]

Throughout the hearings, an antimonopolistic bias against the railroad in-dustry dominated discussions about what to do with the tunnel and railroads that served it. This bias could be traced partly to the Bay State's disappointing experience with the Western Railroad as a conduit for western grain in com-petition with New York State. Speaker after speaker condemned the Western for putting its shareholders before the good of the people and foresaw a similar outcome if the consolidated railroads along the Hoosac route fell under private control. This view also mirrored the national reaction to the railroad indus-try's controversial behavior in the years after the Civil War. Charles Francis Adams, Jr.—the head of the Massachusetts Railroad Commission and author of *Chapters of Erie*, a scathing expose of the railroad industry—set the tone of this broad condemnation at the start of the hearings. While normally against state intervention in commerce, Adams declared railroads were different and "not subject to the laws of supply and demand." The state now found itself in a situation, he continued, where "one branch of business of the community is so monopolized that the citizen can no longer derive the advantages from the benefits of competition." Esteemed figures such as Josiah Quincy, Jr., President Chadbourne of Williams College, and Ex-Governor Claflin echoed this view and called for state ownership of both the Hoosac Tunnel and the consolidated line through it. They argued that state ownership was the only way to defend the public interest from the rapacity of railroad speculators and the narrow interest of stockholders. The most enthusiastic advocate of state control was prominent Boston businessman Edward Crane, who proposed that Massachu-setts spend $54 million to build its own railroad all the way to Lake Ontario. This great line would enable the Bay State to compete for western grain with Cornelius Vanderbilt's New York Central Railroad. Only a few speakers, among them Edward Atkinson, argued against state control. Also a member of the local business community, Atkinson represented a conservative faction dia-metrically opposed to Crane and committed to the absolute separation of the state and business. He maintained that government was incapable of running a complex operation like a railroad and that the tax increases necessary to purchase a consolidated line through the Hoosac Tunnel, or an extension of that line further west, would be disastrous for local businesses. In the end, Bay State legislators were so confused by the debate that they postponed any decision regarding management of the tunnel until its completion. Their indecision was a bad sign for the tunnel's future.[7]

So, what had happened to generate such animosity toward America's railroads? After all, only two decades earlier, Nathaniel Hawthorne and his intellectual cohorts had been enthralled by the coming of the railroads. Hawthorne's description of Clifford Pyncheon and his sisters' first train ride in the 1851 novel *House of Seven Gables* had been rapturous. That same year, Massachusetts legislator and tunnel supporter Whiting Griswold had sanctified railroads as a God-given invention that foretold a new era of peace and prosperity. However, those were the early days of the railroads. There were barely ten thousand miles of railroad track in the nation. That changed as the country grew. By 1860, America boasted over thirty thousand miles of railroad track and, a decade later, over fifty thousand. Not only did the railroad industry keep pace with the nation's booming economy, but it also contributed substantially to it. The growth of railroads drove the country's iron, coal, and manufacturing industries. The railroads also came to dominate America's financial markets. In the years after the Civil War, railroad companies engaged in cut-throat competition and a variety of fraudulent practices. Stock manipulation, inflated construction costs, bribery of public officials, and discriminatory rate setting corrupted the railroad industry and turned the public against it. The struggle for control of the Erie Railroad, chronicled by Charles Francis Adams, Jr., in *Chapters of Erie*, captured the worst of these practices. Cornelius Vanderbilt had taken control of the New York Central Railroad and attempted to do the same with the competing Erie. The ensuing "conflict of Titans" was fought in the New York State court system and on Wall Street. Both Vanderbilt and his rival Daniel Drew were masters of stock watering and other forms of financial legerdemain. Two other speculators, James Fisk and Jay Gould, eventually took control of the Erie, looting it of its assets and bankrupting the company. A second scandal that gained widespread public attention involved Crédit Mobilier, a firm doing inflated construction work for the Union Pacific Railroad. Profits from Crédit Mobilier and stock in the lucrative enterprise were used to buy influence for the railroad in the U.S. Congress. Its exposure during President Grant's 1872 reelection campaign sullied the administration and ended federal funding for the transcontinental railroads. At the center of the scandal was Massachusetts congressman Oakes Ames. One historian has described the post–Civil War years as a period of "degeneration" for the railroad industry.[8]

The Panic of 1873 slammed Massachusetts and the nation during September of that year, just as miners were boring out the last thousand feet of the Hoosac Tunnel. It was a sad coincidence that would have negative ramifications for the tunnel. The precipitating event was the failure of Jay Cooke & Company, a bank Cooke owned and had used to finance his takeover of the Northern Pacific Railroad. In an unprecedented move, the New York Stock Exchange closed for ten days. The crisis caused President Grant to visit Wall Street on a fruitless rescue mission. Within months of the panic, railroad stocks lost 60 percent of their value, more than any other sector of the market. Roughly half the railroads in America would eventually default on their debt. In modern parlance, the panic began as a "railroad bubble." However, it spread through the economy with a speed and scope that stunned the public. Few realized the ensuing depression would last half a decade. It would be the worst economic collapse the country had experienced.[9]

It is impossible to know how many miners at the Hoosac Tunnel knew what was happening to the national economy. The few with access to newspapers and the ability to read them were probably thankful to have jobs. In Boston and New York, mounting bankruptcies were idling workers by the thousands. Still, those places probably seemed distant and irrelevant to miners driving the Hoosac Tunnel toward completion. After all, they were engaged in something momentous.[10]

By mid-November 1873—at a distance of approximately 10,125 feet from the west portal and some 2,043 feet from the central shaft—miners converging on each other knew they were close. For miners at both headings, the pounding of the Burleigh drills working toward each other had grown louder. Both teams could feel the vibrations of each nitroglycerin explosion under their feet. When the two headings were sixteen feet apart, Walter Shanly ordered work stopped. Invitations went out to politicians and dignitaries for the "final blast," scheduled for two o'clock in the afternoon on Thanksgiving Day, November 27, 1873.[11]

On a cold, snowy Thanksgiving Day, around six hundred people converged on the Hoosac Tunnel from different directions. Many from North Adams tramped through the west portal and collected at the west heading. At least two hundred rode up the Hoosac Mountain on sleighs and took the elevator down the central shaft. Once there, they walked west through several hundred feet of fully carved-out tunnel before the going got rough. The tunnel became more confined and was filled with equipment and debris. Other attendees, including Walter Shanly and a dozen or so legislators who had enjoyed lunch at Rice's Hotel, rode the narrow-gauge train two miles from the east portal to the bottom of the central shaft. Some of these dignitaries sat on jagged rubble in the mining cars. From there, they joined the tightly packed crowd at the east heading. The celebrants consisted of worthies from Boston, there to satisfy their curiosity and be part of history, and common miners, referred to locally as "muckers," wanting to see their enormous labors finally consummated. For the latter group, who had seen almost two hundred of their colleagues either killed or severely injured in the great work, the moment must have been poignant. Despite the chill and dripping water inside the tunnel, the mood was ebullient. A group of Williams College students provided vocal entertainment. Several days earlier, miners had drilled a two-inch speaking tube through the remaining rock between the two headings. Professor Mowbray had placed 155 pounds of frozen nitroglycerin in the west heading and ran the detonation wire through the speaking tube. Finally, at the appointed hour, miners backed the celebrants away from both headings and closed the heavy wooden doors that had been installed for the occasion, isolating them from the blast area. In the silence that followed, Walter Shanly pushed the handle down on the detonator, igniting the final blast. The tunnel was plunged into darkness and the air filled with rock dust. The celebrants, cringing behind equipment and boulders, took a moment to recover and then sent up a loud cheer. They relit their lanterns and gazed about. Shanly led the way to the five-and-half-foot round hole joining the two headings. He stopped momentarily to ensure the way was safe. Then, always the gentleman, he ushered state senator Robert Johnson, chairman of the tunnel's joint legislative committee, through the opening ahead of him. Celebrants from both headings exited the west portal together. Wheeler's Brass Band and Drum Corps accompanied them to the Arnold House in North Adams for a massive reception. The revelry continued into the night at the town's opera house. The Hoosac Mountain had been breached.[12]

After living with the on-again, off-again drama of the Hoosac Tunnel, its succession of contractors and engineers, and its frequent setbacks and disasters for nearly a quarter of a century, the residents of North Adams and nearby towns were proud and ebullient. "The great work is assured," the Adams Transcript declared. The local population had endured not only the "croakers" and "naysayers" over the years but also the "conspiracies and jealousies of rival roads." At last, believers in the tunnel were "completely vindicated." It had been an "undertaking unparalleled in the history of the country," the newspaper boasted. "The gigantic mountain barrier that separated the West from the seaboard is conquered and removed, and now the mighty stream of traffic can flow forever over the state to the sea."[13]

However, the final blast and celebratory walkthrough did not mean the Hoosac Tunnel was finished. No golden spike had been driven at the Thanksgiving Day event. Fourteen months would elapse before the first railroad train passed through the tunnel. Enough work remained that Walter Shanly would have to extend the completion date of his contract from March 1 to September 1, 1874. Importantly, the two-and-a-half miles between the west portal and the North Adams depot remained impassable. A deep cut had to be made through hard, rocky ground and a roadbed graded before tracks could be laid. East of the tunnel, the challenge was less daunting but still critical. The railroad track there was poorly aligned, broken down by time and weather, and too light for the heavier freight expected to use it. The state decided to replace fifty-six-pounds-per-yard iron rails with sixty-eight-pound steel rails. So, too, the Deerfield Bridge needed reinforcing. Perhaps it was inevitable that a railroad which had taken so long to complete would require updating before modern locomotives with their heavier loads could traverse it.[14]

Walter Shanly was responsible for laying the track from the west portal to the North Adams depot but not for the track improvements east of the mountain. However, the greatest threat to his diminishing profit margin was the need for additional arching inside the tunnel. He had already dismantled his brickyard and knew that any new arching required enlarging the tunnel's dimensions to accommodate it. Nonetheless, during January 1874, Consulting Engineer Philbrick ordered Shanly to arch a 1,800-foot section of the tunnel just west of the central shaft. The tunnel's ceiling was unstable there, and falling rock had already cost one miner his life. The additional arching would cost

$300,000. Shanly refused to comply with Philbrick's order because he believed his contract only called for arching in the section closer to the west portal and specified the exact number of bricks to be used. Shanly had complied with these requirements and argued his case several times before the governor and the legislature. Eventually, Shanly prevailed. This was fortunate because the problem of falling rock would grow worse.[15]

There were several theories as to why the tunnel was so unstable. The variability of the mountain's core material—soft, flaky mica-schist mixed in with hard gneiss and quartz, not to mention fragile soprolite, or so-called porridge stone—certainly contributed to the problem. During the winter months, freezing water in the tunnel's ceiling tended to pry some of its more friable material loose. Other experts suggested the mountain was "settling" after so much material had been excavated from it. Still others blamed the use of trinitroglycerin, especially where top-header excavation had been employed. These experts claimed that the explosive was too powerful and its shock waves had damaged the natural structure of the rock above the tunnel's ceiling. Whatever the cause, falling rock was killing miners trying to trim out the tunnel and grade its roadbed and would make train traffic through the tunnel impossible until the problem was fixed. Before it was finished, Massachusetts would be forced to install another 5,599 linear feet of arching to insure the tunnel's integrity. When added to the 1,974 feet B. N. Farren and the Shanlys had constructed, a total of 7,573 linear feet of arching would support almost a third of the tunnel's 25,081-foot length. Yet another of geologist Edward Hitchcock's optimistic predictions from 1848—that the Hoosac Tunnel would require little or no arching—had proven tragically misleading.[16]

Walter Shanly did not bid on this arching work or the track improvements east of the mountain. He wanted to go home. Just days before Christmas 1874, six years after signing his contract with Governor Bullock, Shanly got his wish. He received $456,014 for extra costs incurred and withholding on his monthly payments. This included certain adjustments for extra work and work not completed. Shanly reserved the right to present claims later for money he believed he was still owed. Over the next decade, he would return to Massachusetts numerous times to argue his case. Shanly would be paid $188,000 in 1875, but a further $129,000 approved by the legislature would be vetoed by Governor Benjamin Butler. In the end, Shanly dismissed Massachusetts politicians as a "lot of the most contemptable scalawags" and vowed never to work in the state again.[17]

Just days after Shanly's contract was certified complete, Alvah Crocker died. He had caught a fatal case of pneumonia returning from the nation's capital to Fitchburg. He was seventy-three years old. He was buried in the town's Laurel Hill Cemetery. Although congressional responsibilities had kept him away from the Thanksgiving Day ceremony the year before, Crocker had the chance to walk the length of the tunnel some months later. It must have been exciting for him. Unfortunately, the father of the Hoosac Tunnel would never see a freight train pass through the great barrier that had stood in the way of his dreams of western bounty and renewed prosperity for his state.[18]

In its final act, the tunnel drama's cast of actors changed again. Yet, many of those who took the stage were tunnel veterans. With Walter Shanly's departure, the state needed a more qualified consulting engineer than Edward Philbrick. To fill that role, Thomas Doane returned from Nebraska, where he had served as chief engineer for the Burlington & Missouri River Railroad. Alvah Crocker had fired Doane during an 1867 frenzy of cost-cutting. Working underneath Doane as associate engineer was W. P. Granger, who had quit the tunnel after the central shaft disaster, thinking he might be implicated. Also attending this reunion was B. N. Farren, who would complete the additional arching in the tunnel and upgrade a section of the tracks east of the mountain. So, too, A. E. Bond, a native of North Adams who had spent his entire career at the tunnel, would take on the dangerous job of removing the timber platforms, elevator system, and other equipment from the central shaft. A newcomer, Norman Carmine Munson, would be responsible for most of the track upgrade from the east portal to Shelburne Falls. Munson had spent the previous decade filling in the Back Bay section of Boston. These men would report to a new oversight organization of five "corporators," chosen by the governor. Among the most notable were Charles Francis Adams, Jr., President Chadbourne of Williams College, and Ex-Governor Washburn. (Washburn had taken the U.S. Senate seat of deceased Charles Sumner.) The state allocated $1.5 million to finish the tunnel and bring it into commercial operation.[19]

Again, if laborers at the tunnel knew what was happening in the economy around them, they were happy just to have jobs. And finishing the tunnel was labor-intensive. B. N. Farren's arching alone kept some five hundred workers busy laying up bricks and mortar inside the tunnel. All the while, economic depression ravaged Massachusetts and the nation. Although exact numbers are difficult to arrive at, it is estimated that the overall economy contracted by roughly 30 percent in the first year of the depression. As bankruptcies soared, wage cuts and layoffs spread. In Boston, idled laborers petitioned Mayor Cobb to create public works programs so they could earn enough to feed their families. Cobb claimed the city had no money to do so, having depleted its resources after the Great Fire a year earlier. In these desperate circumstances, the city's better-off reformers founded the "Co-operative Society of Volunteers among the Poor" to help feed starving families. David King, a manufacturer of straw hats in the city, argued in favor of price controls to restore purchasing power and prevent deflation. H. H. Bryant, a wholesaler of clothes in Boston, accused British banks of conspiring to destroy the Yankee economy. Others blamed the collusion of business and government that had allegedly led to monopoly, high tariffs, and financial abuse. Cynicism and disillusionment reigned. President Chadbourne of Williams College declared that the elevation of the masses in America had been "the result of the chance conditions of our new fertile country rather than of the superiority of any system of labor that prevails amongst us." In other words, the free-labor ideology over which the Civil War had been waged was now discredited by what had happened to the economy. Similarly, any deference for the Massachusetts working class—so pridefully espoused by Edward Everett, Elias Hasket Derby, and others in the Puritan bloodline—had all but dissipated. Chadbourne went on to say that Boston and other cities should expel any man "that cannot show honest means of living."[20]

Already weakened by scandal and factionalism, the Republican Party staggered from the impact of the depression both nationally and in Massachusetts. In the congressional elections of 1874, the Republicans went from a 70 percent majority in the U.S. House to a 37 percent minority. (The Republicans held on to the U.S. Senate because few seats were up for reelection.) In Massachusetts, the party was beset by disillusionment and division. William Robinson, a former Bird-man, described the party as "an old and battered umbrella, of no use for shelter to anyone and yet impossible to shut." After President Grant's mishandling of local patronage and Governor Washburn's ham-handed application of state liquor laws, the Massachusetts Republican Party began to hemorrhage. Many once-loyal party members migrated first to the Liberal Republicans, then to a fusion ticket with the Democrats, and finally to the Democrats themselves. Unthinkably, Francis Bird led the way. The Panic of 1873 and the Republicans' inability to address the hardships it imposed on Bay State voters broke the party's back. In the election of 1874, the Democrats took control of the Massachusetts legislature and won the governorship. William Gaston, mayor of Boston during the Great Fire, defeated Thomas Talbot, Governor Washburn's handpicked successor. Massachusetts had not elected a Democrat governor in a quarter of a century. Still, many voters saw the two parties as indistinguishable and equally feckless. Voter apathy reduced turnout to a record low. Mired in a grinding depression and doubting many of the values that had inspired them in the past, the citizens of the Bay State now turned their attention to the question of what to do with the Hoosac Tunnel.[21]

The lifeblood of the Hoosac Tunnel had been both material and spiritual. More specifically, the tunnel had been nourished by state funding and public belief in its economic promise. State funding was tenuous during the tunnel's early years, but increased substantially after the state took it over. Funding only flowed in abundance once the tunnel reached a certain stage of completion and opposition to it collapsed. Public belief in the tunnel's economic promise is harder to quantify. Nor can the contours of that belief and how it changed over time be easily described. Apparently, blind faith in the tunnel's economic promise and its gradual politicization simplified the tunnel's proposition such that hard questions about its ultimate utility were never asked. It was as if the mists of the Hoosac Mountain and the fog of Boston Harbor had combined to mask the tunnel's insufficiency as a stand-alone project and obscure the lack of a connecting infrastructure to make it part of a viable transportation system. How exactly was western grain going to reach the tunnel, and, if it did, how was that grain to be dealt with in Boston and then transported to Europe? In short, the Hoosac Tunnel was a chimera—a strange creature lavishly draped in western bounty but mute as to how that bounty would find its way to the tunnel and then to foreign markets.[22]

Though naïve and poorly articulated, the tunnel's economic promise had contained its share of magic. In the unique historical moment of its conception, dreamy romantic notions and exciting technical advances polished its allure. The Hoosac Tunnel scheme was animated by the lure of western agricultural abundance and the transforming technology of the railroad. Protunnel journalists never ceased stimulating the public imagination with visions of economic renewal through "a more intimate intercourse between the West and Boston." Overflowing with wheat and other foodstuffs, the West was like an immense national warehouse waiting to be drawn on by eastern port cities for export to a hungry Europe. Given the seemingly limitless bounty of the West and what it had done for New York, the flow of commerce through the Hoosac Tunnel was expected to not just improve the Bay State's economy but to transform it. The miracle of the railroad made it all seem feasible. Respected intellectuals such as Ralph Waldo Emerson swooned before the railroad's transformative potential: "Railroad's iron is a magician's rod, in its power to evoke the sleeping energies of land and water." In the thrall of so much technical progress, everything seemed possible. And yet, neither the mystical West nor the iron pathways from it to Boston were clearly defined. They floated like points and lines on a schoolboy's map, captivating yet imprecise.[23]

Perhaps if it had not taken so long to complete the Hoosac Tunnel might have turned out differently. After all, Massachusetts financiers such as Nathaniel Thayer, John Murray Forbes, and James Joy had been active in western railroad building during the 1850s and could have played a role in extending the tunnel route to the West. These investors had been the financial backbone of the Michigan Central Railroad, for example. "They had built great railroads throughout the west," according to Charles Francis Adams, Jr., "but those roads did not lead to Boston." Had these investors purchased the New York Central Railroad when they had the chance—the northernmost "trunk" line connecting Albany and Buffalo—then Massachusetts would have captured the optimal connection between the tunnel in the northwestern corner of the state and the American heartland. (The other three trunk lines to the West—the Erie, the Pennsylvania, and the Baltimore & Ohio—all ran too far south to connect efficiently with the tunnel.) Instead, New York railroad speculator Cornelius Vanderbilt, or the "Commodore" as he was popularly known, took control of the New York Central in 1867. He quickly merged it with his Hudson River Railroad running from Albany to New York City. To this 740-mile behemoth, Vanderbilt soon added the Lake Shore Railroad, connecting Buffalo and Chicago. The bulk of American wheat—by now the nation's most important export—went mostly by rail from western farms to Chicago before traveling to the eastern seaboard. By 1870, Vanderbilt was shipping over four million tons of western freight from Chicago to the Port of New York over almost 1,200 miles of continuous rail line. Given his desire to maximize profit and efficiency, Vanderbilt saw no reason to divert much of this freight to Boston. Nor did Massachusetts have a way to compete with Vanderbilt's railroad. As a consequence, the Bay State found itself more isolated than ever. Time had caught up with the Hoosac Tunnel.[24]

Even had it had the chance, it is doubtful whether Boston could have handled western commerce of any magnitude in the 1870s. In addition to its natural disadvantages versus New York Harbor, the Port of Boston had never invested enough in its infrastructure. Western freight arriving in New York, on the other hand, went directly by rail to facilities designed to deal with large volumes. The finest was St. John's Park on Manhattan's Lower West Side, a massive city-within-a-city of grain depots, warehouses, and stockyards. New York's storage capacity was a magnet for arriving ships. Their captains knew they could easily pick up return cargo, which was not the case in Boston. Also, New York was a relatively inexpensive port of call. Deep-water wharves lined the entirety of Lower Manhattan. This made wharfage rates for ships unloading and loading goods a fraction of what Boston charged. Although Boston had promised its western visitors in 1865 to address the inferiority of its port facilities and numerous governors had campaigned on the issue, the city had failed to modernize its harbor side and expand its warehousing capacity to serve as a major *entrepôt* for western goods.[25]

Similarly, Boston's ocean link to Europe had deteriorated over the years. Emerson had written that ocean-going steamships had "narrowed the Atlantic to a strait." But Massachusetts shippers did not heed him and retreated before the challenge of steam navigation. Part of the problem traced to opportunities that encouraged an adherence to sail. These included a lucrative packet-ship business along New England's northern coast, the demand for fast clipper ships during the California Gold Rush of the late 1840s, and a boom in Canadian trade after a reciprocity treaty was signed in 1854. Another problem had to do with the Bay State's disadvantaged coastal geography. There were few protected bodies of water to serve as "steamboat nurseries," the way they did in New York State. It was in New York Harbor, on the Hudson River, and along Long Island Sound that entrepreneurs like George Law, Elisha Peck, and Cornelius Vanderbilt perfected their steamboats during the 1830s. Responding to demand for southern cotton, they established routes to Charleston and New Orleans during the next decade. Just prior to the Civil War, New York steamboat magnates like Vanderbilt and Edward Collins connected to Europe with their large wooden side-wheelers. Massachusetts lagged this development, with no steamboat line south of Cape Cod until the 1850s and none to New York until after the Civil War. Following that conflict, steam navigation came of age with the introduction of iron hulls and screw propulsion. However, the British dominated transatlantic navigation with their faster, more efficient fleets. Still, British shippers favored New York for its supply of cotton, vital for their country's textile mills, and its ample stores of western wheat. When the price of American wheat sank low enough to compete with European wheat, New York wheat exports burgeoned. For its part, Massachusetts' ostrich-like attitude toward steam navigation bore bitter fruit. The state had little to attract foreign steamships or sustain local lines. The British Cunard line diverted its mail steamers from Boston to New York in 1868. A year later, the American Steamship Company's Boston-Liverpool route failed, taking its Boston stockholders down with it. In 1872, another British company, the Inman line, chose to make Boston a port of call on the way to New York. It pioneered its service with the steamer *City of Boston*, which on its return to Liverpool was never heard of again. Inman ceased its Boston visits.[26]

"The Hoosac Tunnel [was] a monument of the energy and perseverance of Massachusetts rather than of its wisdom," a state senator admitted in a moment of candor. So, how were tunnel supporters able to ignore the broader commercial framework necessary to access the granaries of the West and deliver their bounty to a hungry Europe? How did the symbolism surrounding the tunnel obscure the bigger picture of what was required to do this? Without a rail connection to the West, an adequate infrastructure at the Port of Boston, and a steamship link to Europe—the tunnel stood no chance of rejuvenating the Bay State economy. Possibly, this blindness to reality had something to do with the way Massachusetts had revived its faltering economy in the past. All those instances of innovation shared a kind of serendipitous, random-walk quality about them. For example, the great experimentalist Elias Hasket Derby had continuously probed for new opportunities with his merchant fleet in the Far East, Baltic states, and other parts of the world. He was never sure what he could sell and only established his triangular trading schemes when something panned out for him. In a way, Francis Cabot Lowell was an experimentalist too. He tried out various products in the marketplace before scaling up his production lines behind a best-seller. Once he identified his product offering, nobody could match his efficiency and quality. So, too, Massachusetts whalers relentlessly sought out new hunting grounds around the globe ahead of less-adventuresome rivals. Their self-contained factory ships allowed them a huge range in their hunt. All these endeavors required fluid global economy and a highly flexible approach on the part of these innovators to try out new things and then refine their business models in times of crisis.[27]

But economic life in the mid-nineteenth century changed. The American nation and the world were more interconnected than they had ever been. After the Civil War, the American economy was less flexible, more fixed. Business corporations became much larger. Ironically, the railroad—one of the magical components of the Hoosac Tunnel dream—was responsible for this development. Railroads were bigger and more complex than any type of business before them. Building railroads required more capital than launching merchant ships or erecting factories. While Boston financiers invested in various railroads inside and outside their state, they did not participate in the construction of the great trunk lines to the West. Nor did they attempt to build far-flung railroad empires that would benefit their city. For the most part, Boston money men preferred smaller, more closely held corporate entities. Men like John Murray Forbes and Nathaniel Thayer were members of an elite Brahmin social circle, the former descended from a merchant dynasty involved in the China trade and the latter with roots back to the Puritan founders. They were conservative and did not trust outsiders. Their more modest corporations were perfectly suited to manufacturing, whether textiles, leather goods, or paper. New York's money culture was different. Although New York had its merchant elite, there were also men like Daniel Drew and Cornelius Vanderbilt, the former starting life as cattle drover and the latter as a farm boy. Another was the infamous Jay Gould, who also began life as a farm boy, turned blacksmith, surveyor, and eventually railroad magnate. Such rough, "go-ahead" types lacked social connections but were wildly ambitious and brutally competitive. Once they learned how to leverage the immense capital available to them in the stock market, investing in far-flung steamboat lines and giant railroad enterprises gave them little pause. In the end, Massachusetts continued to lead New York

State in manufacturing. However, New York City became the nation's preeminent commercial center. If the Hoosac Tunnel was a monument to anything, it was to the futility of challenging New York City's commercial ascendancy in a significant way.[28]

The first train passed through the Hoosac Tunnel on February 9, 1875. The eight-wheeler *N. C. Munson,* whose owner was renovating the roadbed east of the tunnel, pulled more than a hundred guests in a boxcar and on two flatcars. Squeezed into the locomotive's cabin were Thomas Doane, B. N. Farren, W. P. Granger, N. C. Munson, plus various other guests and the engine crew. Entering the east portal at three o'clock in the afternoon, the four-car train emerged on the opposite side of the mountain in just over a half hour. However, little fanfare attended this inaugural passage. The weather was bitterly cold, and the participants were probably not sure what they were celebrating. The single track through the mountain had not been properly "balanced," its ties laid temporarily on the tunnel's uneven floor for a slow, rough ride. Nor was the tunnel's arching complete. A wag writing for the local newspaper warned travelers to have their insurance paid up before attempting such a trip. Nonetheless, in early April 1875, the *Deerfield* hauled the first freight train through the tunnel. It consisted of twenty carloads of grain from Troy bound for Fitchburg. By that time, the floor of the tunnel had been graded and the tracks balanced. Still, the train had to be pushed from the west portal uphill to the crest at the central shaft by the *Robinson.* It then coasted down and out the east portal by itself. The train made it through the tunnel in forty-five minutes but was delayed four days at Bardwell's Bridge, which had been washed away by a recent flood. The experience called attention to the work still needed to make the Hoosac line fully operational. A month later, forty tons of the tunnel's ceiling collapsed, crushing a work train. Miraculously, there were no fatalities and no negative publicity to sully the fledgling operation. In October, Governor Gaston rode through the tunnel with his State House entourage. The locomotive was renamed in honor of the governor, and his portrait hung beneath its head-lamp. This time, there was more bunting, more speeches, more lemonade and liquor.[29]

And, yet, the gubernatorial visit was not the happy event it might have been. While the tunnel was operational, it was not so in a way that many tunnel supporters had anticipated. Those who harbored grand hopes for the tunnel were deeply disappointed by what had occurred in the State House during the first months of 1875. The tunnel debate that took place there was similar to the one two years earlier. The legislature was still at loggerheads over whether to consolidate railroads along the tunnel route and whether to extend it to the West. Again, legislators debated the question of state ownership versus privatization. Still, careful observers of the two debates noticed important changes. Certain attitudes within the legislature had hardened and now came to bear on the fate of the tunnel. There was greater fear than ever that Cornelius Vanderbilt and his New York railroad clique were seeking control of the tunnel. As a result of this fear, the legislature had refused to cooperate with a group of Boston investors trying to establish a railroad from the Hoosac Tunnel to Schenectady, New York. Suspecting the new line was a Trojan horse for Cornelius Vanderbilt, the legislature refused to support it. Most important, the legislature had become more fiscally conservative. With the cost of finishing the tunnel mounting and tax receipts reduced by economic depression, the legislature had little stomach for new outlays behind the state's infrastructure. Nonetheless, Edward Crane was back at the State House with his plan for a trunk line from the Massachusetts border to Lake Ontario, running north of Vanderbilt's New York Central Railroad. This time, he estimated the cost at $35 million. But fiscal retrenchment was in the air, and Crane was shouted down. It was at this point that a little-known representative from Somerville, S. Z. Bowman, rose from his seat and offered his toll-gate plan. This plan called for the state to retain ownership of the tunnel but make it available to any railroad wanting to use it. Any railroad could "enter upon and use" the tunnel for "just and reasonable tolls." No consolidation of the railroads on the tunnel line would be necessary under such a plan, and the tunnel would be protected from the threat of an out-of-state takeover. It was also the lowest cost option of any plan before the legislature. Much to the horror of Charles Francis Adams, Jr., Edward Crane, and others with loftier visions for the tunnel, "Mr. Bowman's Bill" passed the legislature and was signed into law during April 1875.[30]

Some who voted for the toll-gate plan did so because they thought it would be temporary. It would be a kind of "experiment" and, according to one legislator, "will not allow the interest of the State in the tunnel to be disposed of before we know its value." Others felt the toll-gate plan would keep the tunnel out of the hands of powerful railroad interests, both inside and outside of

Massachusetts, and thereby allow it to better serve the public. "It is of great merit that this bill does not provide for the creation of a great consolidated corporation," one toll-gate supporter commented. Still others who had been earlier boosters of the tunnel had since come to believe it was no longer viable. After all, Elias Hasket Derby had called the legislators' attention during the 1873 debate to the lack of a connecting trunk line to the West and the inadequacies of the Port of Boston. In spite of these problems, a substantial minority in the State House saw the toll-gate plan as impractical and voted against it. The *Springfield Republican* compared it to an antiquated village common, "accessible to all passersby and everyone's cow but the very antipode of successful farming." Not only did the toll-gate plan discourage consolidating the various railroads along the tunnel route, but it posed an impediment to any future opportunity for a Massachusetts-owned trunk line to the West. Importantly, the tolls to be collected under the plan would never be enough to cover interest payments on the outstanding tunnel debt. In a bizarre twist of fate, Massachusetts drew back at its moment of triumph. Shaken by economic depression and intimidated by the power of its great commercial rival, the state opted for frugality and the protection of its unique gateway to the West.[31]

In the wake of the toll-gate decision, Edward Crane and his Boston business circle continued to lobby for their trunk line to the West. They organized the Boston & Chicago Railway Trust Company to help finance the venture and approached the joint legislative committee on railroads multiple times for state aid. One of their petitions was signed by over two thousand businessmen who supported Crane's plan. But the dye was cast. Politicians in the State House ceased talking about forging a rail link to the West and lauded the toll-gate plan for "cheapening the food and fuel to all classes of our people." Early assessments of the tunnel's economic impact under the toll-gate plan were distorted by the long, deep depression following the Panic of 1873. One report claimed the tunnel had made a "fair showing" attracting western freight but did not quantify what that meant. Nor did it identify the New York Central Railroad as the chief conduit for whatever bounty Boston had garnered. This was fine as long as Bay State shippers and the toll-gate manager conducted their business exclusively with the New York Central. Any flirtations with rival trunk lines like the Erie or Pennsylvania invited the wrath of Cornelius Vanderbilt. Although the come-one-come-all rhetoric of the toll-gate plan sounded wonderful, Massachusetts rail traffic through the tunnel depended on the whims of the Commodore.[32]

Because it did not make financial sense, the toll-gate plan was doomed from the start. The high cost of maintaining a safe tunnel and of upgrading the tracks to and from it would consume a large part of the tunnel's revenue. In addition to further tunnel arching, elaborate facades were built at the west and east portals, in 1874 and 1877 respectively, and double tracking was installed through the tunnel in 1881. As traffic increased in the late 1870s and early 1880s, the tracks east of the tunnel needed straightening and refurbishing. In the early 1880s, a locomotive roundhouse was built in North Adams. Between completion of the Shanly contract at the end of 1874 and the state's last tunnel appropriation in 1884, Massachusetts spent $4,640,000 on the Hoosac Tunnel and the Troy & Greenfield Railroad. The state's financial dilemma was exacerbated by the need to pay half the tunnel's revenue to the railroad companies along the tunnel route. After all, every railroad along that route knew trains traveling through the tunnel needed to traverse their tracks too. Accordingly, they drove tough bargains with the toll-gate manager for their share of the tunnel's profit. Out of this paltry income stream, the state could not even consider paying interest on the debt incurred in the original construction of the tunnel. By 1877, annual interest on the Hoosac Tunnel debt was $3,287,834, which exceeded the state's education budget and had to be serviced out of tax receipts. Given the size of this financial burden, it is remarkable that Massachusetts put up with it for a decade after the toll-gate legislation became law.[33]

In the meantime, both the Massachusetts legislature and public were preoccupied with the ongoing economic depression. In 1877, Governor Alexander Rice, a Republican who had replaced Governor Gaston two years earlier, noted that property values around the state had declined to the 1871 level and were still falling. Because it was so dependent on manufacturing, tight profit margins for factory owners and bankruptcies in the industrial sector had a disproportionate impact on Massachusetts. Wages of unskilled workers sank to levels not seen since before the Civil War. Unemployment was so pervasive throughout the state that large numbers of "tramps" roamed the countryside looking for work. Seen as a threat to society, they were forcibly driven out of many towns and denied access to others. Within the state's large urban population, the well-being of the poor deteriorated terribly. Insufficient diets and epidemics took their toll on the young. Infant mortality rose by nearly 60 percent between the Civil War and the middle years of the depression. Surveying this gloomy scene, commentators in the late 1870s described it as "general stagnation" and "national decay." Given its devastating impact, it is no wonder that the great depression of the 1870s not only distracted the legislature and the public from the tunnel but also dissipated much of the optimism that had nourished its construction.[34]

Because work on the Hoosac Tunnel spanned nearly a quarter of a century and straddled the Civil War, the Bay State polity which broke ground on the tunnel in 1851 was not the same one which celebrated its completion in 1875. During this time, the public psyche shifted in ways that affected perceptions of the tunnel and its promise of economic rejuvenation. Writing in his Berkshire cottage in 1857, Herman Melville signaled the beginnings of this change in his last full novel, *The Confidence-Man*. In it, Melville replaces the triumphalism of *Moby-Dick* with a mood of disillusionment and despair. Gone is the whaling ship *Pequod*, with its multiracial comradery and shared quest. Instead, we have the Mississippi riverboat *Fidele*, its faithless and prevaricating human cargo symbolized by a sign hanging on the boat's barber shop reading "No Trust." It is a "ship of fools," representing a society governed by chicanery and swindle, fraud and deceit. The recent success of the Know Nothing Party in Massachusetts and the economic collapse of 1857 cause Melville to single out the Indian-hater and stock-jobber for ridicule. While the free-labor movement has begun to find its footing in the state, Melville cynically lauds factory automation. Similarly, the novel debunks the era's reformist impulse as "mere dreams and ideals," destined to leave "naught but scorching behind." Melville sounds the death knell for the idealism of the antebellum period, which he dismisses as the age of the "cracked pate."[35]

The "scorching" that Melville could neither have foreseen nor imagined arrived four years later with the Civil War. As one historian has observed, literature abruptly ceased because the horrors of the war were indescribable. After the war, Massachusetts and most of the country became changed places. Many of the beliefs that had led to the war were discredited and gave way to skepticism and cynicism. The growth and deterioration of urban areas with their legions of immigrants and displaced farmers marked a crumbling of the old, familiar order. Frauds and scandals in government and industry dominated the newspapers. In 1875, Henry Ward Beecher, the country's most respected religious figure, became the center of a seamy adultery scandal and highly publicized trial. There was a general sense of moral declension. In politics, the Republican Party turned its back on the plight of ex-slaves and coddled big business. In Massachusetts and elsewhere, the Panic of 1873 and the depression that followed shed doubt on the ideal of free labor. The Great Strike of 1877 made many better-off Americans hostile to working men. That same year, the disputed presidential election of 1876 was settled with a political compromise that effectively ended Reconstruction. It was the ultimate betrayal of freedmen's hopes for rights and equality. The idealism that had helped bring about the war and that caused people to believe in the tunnel became a distant memory. The passionate rhetoric that had inspired both now sounded like a foreign tongue.[36]

This collapse in the national mood permeated most cities and towns. The *Adams Transcript* called attention to the "organized corruption of our legislative bodies," "misgovernment of the cities," and "increased crimes of violence and fraud." The Republican Party was to blame for much of what had gone wrong. "The spell is broken," the newspaper declared and predicted a Democratic revival. But it was the people themselves, the paper argued, that had become debased in their attitudes and behavior: "In the very neighborhoods of splendid churches and theaters and libraries, vice and ignorance has produced a race of Huns fiercer than those who marched with Attila."[37]

Against this dismal backdrop, the people of North Adams were glad to at least have the Hoosac Tunnel completed and the Troy & Greenfield Railroad operating for the town's benefit. The deficiencies of the toll-gate plan did not seem to bother them anymore. After all, the Massachusetts taxpayer was covering the railroad's financial deficiencies. And the tunnel *was* busy. In 1876, over 6,800 passengers traveled through the tunnel. There was a high curiosity factor. However, the trip's appeal to travelers quickly waned due to the problem of thick locomotive smoke in the tunnel. Disappointingly, after so much loss of life and treasure to dig it, the central shaft proved totally ineffective in venting accumulated smoke. It was one more tragic miscalculation having to do with the tunnel. The Hoosac line also carried almost a quarter of a million tons of freight, most of it originating west of the Hudson River. As had been hoped, grain made up most of this freight. The best available figures show Boston's grain exports at 8,957,032 in 1878 versus 79,244,083 for New York City. But, while it also lagged behind Philadelphia and Baltimore, Boston had at least joined the game.[38]

Still, the State of Massachusetts could not bear the financial burden of the toll-gate plan forever. Although the Troy & Greenfield Railroad including the tunnel was only forty-four miles long, it was a maintenance nightmare. There were thirty-eight bridges of twenty-five feet or more in length, many of them spanning rivers prone to seasonal flooding. Arching in the tunnel required continuous repairs, as did the sloping of roadbeds that ran along river banks. Station depots, water tanks, coal buildings, road crossings, and miles of fencing had to be kept up. By the early1880s, almost two hundred employees worked on the railroad and tunnel. In its last year of independent operation, the Troy & Greenfield Railroad had net earnings of $113,347 on total revenue of $383,765. Of that, nearly half had to be paid to railroad companies that controlled the route to and from the tunnel. Finally, the state had had enough. In 1885, Republican governor George Robinson announced to the legislature that the toll-gate plan was a failure. Neither his own party—now intent upon fiscal retrenchment and severing the state's relationship with the railroads—nor the opposition argued against him. The Hoosac Tunnel was for sale.[39]

But the tunnel was not for sale to just anyone. It was strictly off limits to out-of-state investors. After two years of courtship, the Fitchburg Railroad came forward as the tunnel's worthiest suitor. However, altruism played little part in the buyer's motivation. The Fitchburg had waited more than a decade until the State of Massachusetts had spent all the money needed to make the Troy & Greenfield a smooth-running operation. It had also waited for the tunnel to build a respectable level of traffic. In the end, the Fitchburg Railroad acted out of fear that another railroad might buy the tunnel. If the Massachusetts Central Railroad, for example, purchased the Troy & Greenfield, it could reroute tunnel traffic south to Northampton and then east to the wharves of Cambridge. The Massachusetts Central could do this because it had a spur connecting it to the Troy & Greenfield just east of the tunnel. Such a route would bypass the town of Fitchburg completely and cut the Fitchburg Railroad out of the Hoosac route altogether. The Fitchburg Railroad bought the Troy & Greenfield Railroad and the Hoosac Tunnel for $5 million in 4-percent bonds and the same amount in common stock. However, the $5 million in common stock was created as window dressing to impress the public. It was practically worthless. The company's preexisting common stock had been converted to preferred stock. The common stock the state received did not pay a dividend until the preferred stock had earned 4 percent. The state never received a cent from the common stock. The state did receive three seats on the board of the Fitchburg Railroad.[40]

The Fitchburg had to pay low for the Troy & Greenfield Railroad and the Hoosac Tunnel because it needed money to fund its larger plans. Specifically, it set about achieving the consolidation of the Hoosac line that the state had failed to bring about a decade earlier. As the major participant in the toll-gate arrangement, the Fitchburg was keenly aware of the need for consolidation and was best equipped financially to make it happen. The Fitchburg also intended to strengthen the Hoosac route's connection to feeder lines outside Massachusetts. In 1891, it bought the tiny Southern Vermont Railroad from the State of Massachusetts for $175,000 in 4-percent bonds. Because it had already leased the Vermont & Massachusetts Railroad in 1874, the Fitchburg controlled the four railroads from Charlestown, on Boston Harbor, to the Vermont–New York border. Next, it purchased the Troy & Boston Railroad for the generous sum of $4 million in bonds and preferred stock. This completed the Fitchburg's control of the original Hoosac line from Charlestown to Troy, New York. However, the Fitchburg went a step further and bought the Boston, Hoosac Tunnel & Western Railroad for $7 million, paid for with a combination of bonds and stock. (About $2 million was in the same nearly worthless common stock it had used to buy the Troy & Greenfield.) This railroad was the same one the Massachusetts legislature had spurned a decade earlier because it suspected it was a Trojan horse for Cornelius Vanderbilt. It had been financed by a consortium of Boston money men, among them Oliver and F. L. Ames, Elisha Atkins, and General William L. Burt. The New York court system and local railroad interests had vigorously opposed the new railroad, as it built westward into New York. However, it had made it from North Adams, Massachusetts, as far as Schenectady, at the head of New York's Mohawk Valley. Schenectady had become a major nexus for the New York Central and other important railroads from the Mid-Atlantic region and the West. Burt and his associates had argued on multiple occasions before the Massachusetts legislature that Schenectady was where the Hoosac line should logically go. Troy, they argued, was a relic of the bygone canal era. While some Massachusetts residents objected to the high price the Fitchburg paid for these two out-of-state railroads versus the dirt-cheap Troy & Greenfield deal, others understood the importance of securing connections to both Troy and Schenectady beyond their own borders. With ample representation on the Fitchburg's board, the State of Massachusetts obviously approved of these plans too.[41]

With its consolidation and expansion by the Fitchburg Railroad, the original vision for the Hoosac Line began to take shape. An 1882 advertising brochure touted the line as connecting the Atlantic to "all the great railroad systems of the Northern States and Canada." Rates would never be higher than between points west and New York. The Port of Boston was, the brochure reminded potential customers, "a day's steaming nearer England than New York." It was much closer than Philadelphia and Baltimore. A map of Boston showed the newly constructed Hoosac Tunnel Docks, next to the Charlestown Navy Yard, with deep-water access. Behind these docks were recently built warehouses and a grain elevator with a capacity of 600,000 bushels of wheat. The Hoosac Tunnel was finally open for business.[42]

So, what accounted for the changing role of the Fitchburg Railroad from reluctant participant in the Great Debate of 1873 to leader of the Hoosac line's consolidation effort in the late 1880s? Frankly, it was a sign of the times. Between 1880 and 1888, some 425 different railroads, or almost a fourth of the country's independent railroad corporations, were taken over by other railroad companies. Overbuilding, cut-throat competition, the beginnings of state and federal regulation, and recurring economic depressions (another one occurred in 1884–85) winnowed down the number of American railroads. By now a mature industry, the better railroad managements recognized the need to expand their systems for competitive leverage and operating efficiency. Because New England had been a pioneer in railroad building and was so thoroughly tracked-over, consolidation happened there early. However, at less than five hundred miles of track between Charlestown and Schenectady, the Fitchburg Railroad was still not large enough for this new, more competitive era.[43]

In 1900, the Boston & Maine Railroad offered to lease the consolidated Fitchburg Railroad. Since the State of Massachusetts was on the board of the Fitchburg, held the largest share of its capital stock—and, by statute, had to approve the consolidation of any railroads entering Boston—Governor Winthrop Murray Crane exercised tremendous leverage over the transaction. Crane negotiated brilliantly. The Boston & Maine was forced to raise its offer for the state's stock in the Fitchburg from $1 million in cash to one of $5 million in 3-percent gold bonds. Furthermore, the Boston & Maine, as a New Hampshire corporation, had to provide assurances that it was not in league with any New York railroad interests or controlled by them before the Fitchburg lease could be approved. Once these issues were resolved, the marriage of the two railroads seemed a perfect match. The Boston & Maine lacked the kind of access to the West enjoyed by the Fitchburg but could draw on commodities from northern New England to supplement the Fitchburg's westbound freight cars. The Boston & Maine also owned much better facilities on the Boston harbor front than the Fitchburg had. Importantly, the Boston & Maine was a superbly managed railroad, somewhat of an anomaly for its time. Its president, Lucius Tuttle, had built a strong and pristine balance sheet, free of stock-watering and other financial abuses that typified the era. Better still, Tuttle had leased the Boston & Lowell Railroad in 1887 and was on the way to achieving the critical mass needed to survive in a rapidly consolidating industry. The Boston & Maine would have a long corporate life and become closely identified with the Hoosac Tunnel. Like most American railroads, however, it would struggle during the mid-twentieth century under pressure from highway and air transportation. The collapse of New England manufacturing in the 1930s hurt the railroad's freight business. After declaring bankruptcy in 1970, the Boston & Maine reinvented itself several times but vanished from the nation's rail system in 2006. The Pan American Railways Company acquired the Boston & Maine at that time and with it the Hoosac Tunnel. The company owns and operates the tunnel today.[44]

CONCLUSION

SO, HOW SHOULD historians assess the Hoosac Tunnel? Even the tunnel's harshest critics—including leading anti-tunnelite Francis Bird—did not foresee the tunnel's full cost, long years to complete, or number of lives lost in its construction. The tunnel was unreasonably ambitious even in an America convinced of its indomitability versus nature and possessed of an appetite for internal improvements. Although it would ultimately prove an engineering success, the tunnel was a poorly conceived project that suffered from a number of tragic misjudgments. The most serious were Edward Hitchcock's original analysis of the Hoosac Mountain's geological composition and Charles Storrow's insistence on a central shaft. Hitchcock could not have gotten it more wrong. His conclusion that the mountain was an "extremely simple formation" and would be "easy for drilling" convinced the Massachusetts legislature to charter a railroad through one of the highest mountains in the Berkshires. His further predictions that the inside of the mountain would be dry and need little arching were also wrong. Arguably, Storrow's central shaft added 30 percent to the cost of the tunnel and sacrificed many miners to a worthless endeavor. It never vented locomotive smoke from the tunnel, nor did it provide additional headings in time to significantly accelerate the penetration of the mountain. Other mistakes included the construction of the Deerfield Dam complex and the deployment of massive boring machines versus the more timely development of the lighter, pneumatic drills. Some of these errors can be laid at the feet of the tunnel's early contractors. However, the real problem was that the mining technology needed to penetrate the Hoosac Tunnel simply did not exist until after the Civil War, a decade and a half after miners had broken ground on the tunnel.[1]

During the final decade of work on the tunnel, Walter Shanly benefited from tunneling innovations such as the Burleigh drill, tri-nitroglycerin, and electrically ignited fuses. Still, postwar inflation and a seemingly endless list of requirements to make the tunnel operational drove the cost of the tunnel to staggering heights. By the time the State of Massachusetts had sold off its last tunnel-related securities in 1897, the total cost of the tunnel including principal and interest came to $28,856,396 (equivalent to $875 million in 2018 dollars). Once contributions to the sinking funds and purchase payments from the Fitchburg Railroad and Boston & Maine Railroad were subtracted, Massachusetts taxpayers were on the hook for around $10 million (over $300 million in current dollars). Boston businessman Edward Atkinson never tired of calculating what the tunnel had cost the average working man. Those who believed the Hoosac Tunnel should never have been taken up in the first place recalled the advice of Ansel Phelps, an attorney for the Western Railroad, who suggested that slabs be erected at the tunnel's two portals declaring the project a "Monument to the folly of Massachusetts."[2]

Still, just as it is an understatement to say it was over budget, it is an overstatement to call the Hoosac Tunnel a "folly." During the 1880s, the Hoosac line became an important freight route for western goods coming into Massachusetts. Though he did not live to see it, Alvah Crocker's goal of transforming the northern tier of the state into a viable commercial corridor became a reality. Before the tunnel was completed in 1875, the Fitchburg Railroad carried 700,000 tons of freight between towns within Massachusetts versus only eight thousand tons to and from towns outside the state. By the early 1880s, however, the Fitchburg Railroad had become the primary conduit for western commodities to the Port of Boston. In 1881, Boston exported nineteen million bushels of western wheat, flour, and corn with a value in excess of $70 million. Over two-thirds of these goods came through the tunnel and down the Fitchburg line to Boston.[3]

In the end, the Hoosac Tunnel revitalized the Port of Boston. By the beginning of the twentieth century—when the Boston & Maine Railroad purchased the Fitchburg—Boston had increased its share of western wheat, flour, and corn exports among the four largest Atlantic ports from 6 percent after the Civil War to 14 percent. From all but New York, Boston picked up a share of western exports. This availability of exportable commodities brought the

Cunard steamship line back to Boston during the 1880s. The Warren and Leyland lines also made Boston a port of call. By 1895, at least one foreign vessel departed each week, exceeding the clearances of Philadelphia and Baltimore. Boston's success was all the more impressive given that it lacked its own trunk line to the West. However, this became less of a problem as the American railroad industry became more competitive and various trunk lines actively sought Hoosac freight. By the 1880s, the New York Central no longer dominated this traffic. After 1887, a newly established Interstate Commerce Commission encouraged a greater degree of cooperation between railroad companies and discouraged rate pooling and other anti-competitive practices. In addition to overcapacity in the American railroad industry, Canada's Grand Trunk Railroad served as a depressant on freight rates. Although it ran north of the Great Lakes and followed a round-about path into New England, this railroad became a notorious freight-rate slasher and forced other east-west lines to match its prices. New England shippers were thankful for the Grand Trunk, glad to be free of the New York Central's domination, and delighted to bargain for freight rates among the various railroads able to connect with the Boston & Maine Railroad and its tunnel.[4]

Ironically, the Hoosac Tunnel never delivered the lower freight-rate differentials that Alvah Crocker and Elias Hasket Derby had so energetically touted before the Massachusetts legislature. In spite of the endless citation of the tunnel route's more favorable gradients, gentler curvatures, and shorter distances versus the Western Railroad and other competing lines, these purported advantages never materialized in a measurable way. Many of these statistics had contained a large component of voodoo. In fact, the freight-rate differential of the Hoosac route remained high even as traffic on it increased. Part of the problem had to do with the heavy debt burden of both the Fitchburg Railroad and, after its purchase of the Fitchburg, the Boston & Maine Railroad. What did make the tunnel line relatively efficient was the large quantity of goods Boston manufacturers shipped westward. Like their rivals in New York, the Bay State's industries sent large quantities of finished goods to western markets in exchange for their wheat and flour. This symbiosis between New England and the West had been the fond hope of Crocker, Derby, and other ardent tunnelites.[5]

And, yet, in spite of what the Hoosac Tunnel did for the Port of Boston and the northern tier of Massachusetts, it never lived up to the overheated rhetoric and grandiose promises that had been used to sell it to the public. The tunnel never seriously challenged the commercial dominance of New York City. Whereas the Bay State's exports had exceeded those of New York at the end of the eighteenth century, they were less than a fifth of its archrival by the 1880s. Nor did the tunnel usher in a new economic era for Massachusetts. Rather, it allowed the state's manufacturing sector to more efficiently reach new markets in the West and restored a measure of growth to Boston's modest export volume. The soaring cost of the tunnel had seemed irrelevant when placed beside dreams of comprehensive economic renewal. What the tunnel contributed to the Bay State economy was impressive but less transformational than its supporters had promised.[6]

And what of the physical tunnel itself? How should history regard it? After all, the Hoosac Tunnel ceased to qualify as the longest tunnel in the Western Hemisphere in 1916. (The five-mile Connaught Tunnel in the Canadian Rockies claimed that distinction.) Today, the length of the Hoosac Tunnel is exceeded by four tunnels in North America. However, it is the earliness of its construction that accords the Hoosac Tunnel its greatness. Because it broke ground in the 1850s and took almost a quarter-century to complete, the Hoosac Tunnel ranks first in audacity, perseverance, and innovation. Initially attacking the nearly five-mile-wide mountain with hammers and hand drills has no rival for audacity. So, too, the perseverance of the tunnel's various contractors and miners who worked on the tunnel remains unchallenged. Without Herman Haupt's perseverance to the point of financial ruin, the tunnel would likely have been abandoned before the Civil War. Without the mining innovations of the Thomas Doane era, the tunnel would never have been finished after the war. All later tunnels owe their existence to the native ingenuity and pioneering technology first deployed at the Hoosac Tunnel. The nearly two hundred casualties that occurred while digging the tunnel attest to the project's difficulty, leaving behind a tragic legacy. In 1927, on the fiftieth anniversary of the tunnel's completion, the *North Adams Transcript* interviewed John W. McManama, one of the so-called muckers who had worked in the tunnel before it opened. "Although men were frequently injured or killed," he remembered, "I kept at work. Many sad things happened but I have tried to forget them." The human cost of the Hoosac Tunnel was extraordinary and begs for commemoration.[7]

CONCLUSION

Today the Hoosac Tunnel is difficult to find. The west portal in North Adams is not accessible to the public and is difficult to visit. The east portal can be reached by an unmarked country road on the other side of the Hoosac Mountain. A few buildings from the tunnel's original construction site can be made out in the woods next to the Deerfield River. They are overgrown with vines and saplings. Several trains pass through the tunnel each day. Gaining a sense of the tunnel itself is not easy. It requires more imagination than looking at the Brooklyn Bridge, for example. The Brooklyn Bridge hangs in the air above New York's East River. At night it is sparkly and reflected in the water below. The tunnel, on the other hand, is buried beneath the weight of the Hoosac Mountain and is pitch black inside. One must drive up the steep east side of the mountain to the tiny town of Florida, more than a thousand feet above the line of the tunnel, and look out from that height over the rolling hills of eastern Massachusetts. Only then can one sense the mass of rock and earth below one's feet through which the tunnel runs. Driving down the sharply curved road on the west side of the mountain into North Adams, one gains a feeling for the four-and-three-quarter-mile distance bored out by miners some century and a half ago. In spite of what one knows about it, the traveler is left with a sense of the tunnel's inherent improbability. Like a blind person running their hands over the sculptured face of Lincoln, one must sense the tunnel's magnificence without actually beholding it.

Perhaps the Hoosac Tunnel deserves to be forgotten. But, by doing so, Americans would also forget the notions that spawned it. The promise of the railroads was in some ways the precursor of our more modern infatuation with technology, so replete with hope one day and disappointing the next. While easterners no longer stand in awe of the West, most do not deny its attraction. Its promises still beckon, even though its politics and lifestyles at times bewilder. In some sense, the notions that produced the Hoosac Tunnel still resonate today, as does the enduring faith Americans place in their grand endeavors.

NOTES

INTRODUCTION

1. *The Writings of Herman Melville*, Vol. 14, 23 (Evanston: Northwestern UP, 1993); and Jay Leyda, *The Melville Log: A Documentary Life of Herman Melville, 1819–1891* (New York: Harcourt & Brace, 1951), 412.

2. Harry Levine, *The Power of Darkness: Hawthorne, Poe, Melville* (Athens: Ohio UP, 1958), 211–212; and Leo Marx, *The Machine in the Garden: Technology and the Pastoral Ideal in America* (New York: Oxford UP, 1964), 170–171, and 286–295. Thomas Carlyle made his arguments in an essay entitled "The Signs of the Times," appearing in the June 1829 edition of the *Edinburgh Review.*

3. Levine, *Power of Darkness*, 217; and Herman Melville, *Moby-Dick; or, The Whale* (1851; rpr. New York: Putnam, 1951), 157.

4. John Kasson has argued that most Americans saw their vast landscape and technical progress as complementary in their sublimity. See John F. Kasson, *Civilizing the Machine: Technology and Republican Values in America, 1776–1900* (New York: Penguin, 1976), 174; Melville, *Moby-Dick*, 158; and Jean V. Matthews, *Toward A New Society: American Thought and Culture, 1800–1830* (Boston: Twayne, 1991), 127.

5. This writer believes Melville knew something about the Hoosac Tunnel. After all, his words "iron rails . . . grooved to run, over unsound gorges, through the hearts of mountains" come very close to describing the tunnel. See Melville, *Moby-Dick*, 157; and Orson Dalrymple, *The History of the Hoosac Tunnel* (North Adams, MA: North Adams Historical Society, 1880), 3, foldout of tunnel profile.

6. Henry Nash Smith, *Virgin Land: The American West as Symbol and Myth* (Cambridge, MA: Harvard UP, 1950), 186; Christian Wolmar, *The Great Railroad Revolution: The History of Trains in America* (New York: Public Affairs, 2012), 35–37; and Peter L. Bernstein, *Wedding of the Waters: The Erie Canal and the Making of a Great Nation* (New York: W. W. Norton, 2005), 127–129.

7. Michael Kammen, *Mystic Chords of Memory: The Transformation of Tradition in American Culture* (New York: Vintage, 1993), 9, and 135–136; and Ronald P. Formisano, *The Transformation of Political Culture: Massachusetts Politics, 1790s–1840s* (New York: Oxford UP, 1983), 173.

8. The Brooklyn Bridge was begun in 1870 and opened in 1883. Its cost has been estimated at

$15 million. About twenty lives were lost in its construction. According to North Adams, Massachusetts, historian Isaac Browne, there were approximately 195 "casualties" at the tunnel, most of which ended in death but some in crippling or maiming. Charles Cahoon, president of the North Adams Historical Society, has documented 135 deaths from newspaper articles and vital records in the towns of North Adams and Florida, Massachusetts. The death toll may have been higher because subsequent deaths caused by injuries were probably underreported locally, especially if those injured moved away. By the time Massachusetts had sold off the last securities related to the tunnel in 1898, according to historian Edward Chase Kirkland, the tunnel's total cost stood at $28 million. See David McCullough, *The Great Bridge* (New York: Simon & Schuster, 1972), 506, 509, 547, and 564; Edward Chase Kirkland, *Men, Cities, and Transportation: A Study in New England History, 1820–1900* (Cambridge, MA: Harvard UP, 1949), Vol. 1, 426; Isaac S. Browne, *Hoosac Tunnel Days: Accidents and Accident Victims, 1859–1878* (North Adams, MA, no publisher and undated); and Charles Cahoon, "Hoosac Tunnel Accidents Compiled by Charles Cahoon" (North Adams, MA: no publisher, 2015).

9. Samuel Eliot Morison, *The Maritime History of Massachusetts, 1783–1860* (Boston: Houghton Mifflin, 1921), 6, 52, and 266; Hamilton Andrews Hill, *Boston Trade and Commerce for Forty Years, 1844–1884* (Boston: T. R. Marvin, 1884), 6–10; and Douglass C. North, *The Economic Growth of the United States, 1790–1860* (New York: W. W. Norton, 1966), 67, 74, 103, and 193.

10. Kirkland, *Men, Cities, and Transportation*, 430–432.

11. William Bond Wheelwright, *The Life and Times of Alvah Crocker* (Boston: Walton, 1923); Francis A. Lord, *Lincoln's Railroad Man: Herman Haupt* (Rutherford, NJ: Fairleigh-Dickinson UP, 1969); and American National Biography Online www.amb.org.

12. Dennis Karwatka, "The Hoosac Tunnel," in *Tech Directions*, 67.9 (Apr., 2008): 9.

13. Louis Menand, *The Metaphysical Club: A Story of Ideas in America* (New York: Farrar, Straus & Giroux, 2001), ix and xii; Jackson Lears, *Rebirth of a Nation: The Making of Modern America, 1877–1920* (New York: Harper Perennial, 2009), 54–56; and Samuel Reznick, *Business Depressions and Financial Panics: Essays in American Business and Economic History* (Westport, CT: Greenwood, 1971), 130, 132, and 136–137.

1. THE LIMITS OF THE BAY COLONY

1. The best summaries of Massachusetts' geological development are at Wikipedia and in a privately published article, "The Geology of Massachusetts," by Patrick J. Barosh and Bradford A. Miller. See https://en.wikipedia.org/w/index.php?title=Geology_of_Massachusetts&oldid=708172177; and www.aegne.org/pdfs/mass_geology.pdf, respectively. See also *The WPA Guide to Massachusetts* (Boston: Houghton Mifflin, 1937; rpr. New York: Pantheon Books, 1983), 9–13.

2. *WPA Guide*, 10–11, and Morison, *Maritime History*, 12.

3. Morison, *Maritime History*, 3–6; and Benjamin W. Labaree, "The Making of an Empire: Boston and Essex County, 1790–1850," in Conrad Edick Wright and Katheryn P. Viens, eds., *En-*

trepreneurs: The Boston Business Community, 1700–1850 (Boston: Massachusetts Historical Society, 1997), 345–346.

4. Alan Taylor, *American Colonies: The Settling of North America* (New York: Penguin, 2001), 175.

5. William Cronon, *Changes in the Land: Indians, Colonists, and the Ecology of New England* (New York: Hill & Wang, 1983), 34–37; and David Cressy, *Coming Over: Migration and Communication between England and New England in the Seventeenth Century* (New York: Cambridge UP, 1987), 2–11.

6. J. H. Elliot, *Empires of the Atlantic World: Britain and Spain in America 1492–1830* (New Haven: Yale UP, 2006), 188–189; Bernard Bailyn, *The Barbarous Years: The Peopling of British North America and the Conflict of Civilizations, 1600–1675* (New York: Knopf, 2012), 329, 365–373; Edmund S. Morgan, *The Puritan Dilemma: The Story of John Winthrop* (New York: Longman, 2007), 48–49; Andrew Delbanco, *The Puritan Ordeal* (Cambridge, MA: Scribner, 1989), 72–73; and Taylor, *American Colonies*, 165–169.

7. Bailyn, *Barbarous Years*, 372–377; Cressy, *Coming Over*, 46–50, 64–66; Taylor, *American Colonies*, 166–167; and E. Digby Baltzell, *Puritan Boston and Quaker Philadelphia* (New York: Free Press, 1979), 95, 110, 125.

8. Delbanco, *Puritan Ordeal*, 34–37; Morgan, *Puritan Dilemma*, 5; and Bailyn, *Barbarous Years*, 390.

9. Perry Miller, *The New England Mind: The Seventeenth Century* (Boston: Houghton Mifflin, 1954), 394–397; Thomas Jefferson Wertenbaker, *The Puritan Oligarchy: The Founding of American Civilization* (New York: Scribner, 1947), 65–67; Morgan, *Puritan Dilemma*, 72–75; and Taylor, *American Colonies*, 185–186.

10. Charles Sellers, *The Market Revolution: Jacksonian America, 1815–1846* (New York: Oxford UP, 1991), 29; David D. Hall, *A Reforming People: Puritanism and the Transformation of Public Life in New England* (New York: Knopf, 2011), 63–69, 129–133; Delbanco, *Puritan Ordeal*, 21–22, 60–61, 126; Wertenbaker, *Puritan Oligarchy*, 69–70; Taylor, *American Colonies*, 170–171; and Bailyn, *Barbarous Years*, 400–401.

11. Mark Peterson has identified "the homogeneity of its population's ethnic origins, religious commitments, and cultural values" as the "most striking feature" of Puritan Massachusetts. See Mark Peterson, *The City-State of Boston: The Rise and Fall of an Atlantic Power, 1630–1865* (Princeton, NJ: Princeton UP, 2019), 579–581; Christine Leigh Heyrman, *Commerce and Culture: The Maritime Communities of Colonial Massachusetts, 1690–1750* (New York: W. W. Norton, 1984), 8–9, 30; Cressy, *Coming Over*, 19; and Delbanco, *Puritan Ordeal*, 8.

12. Heyrman, *Commerce and Culture*, 38–39, 209; and Morison, *Maritime History*, 11–12.

13. Morison, *Maritime History*, 12–15; and Cronon, *Changes in the Land*, 109.

14. Baiyln, *Barbarous Years*, 489, 493–494; and Taylor, *American Colonies*, 185–186.

15. Morison, *Maritime History*, 14–16; Taylor, *American Colonies*, 174–175; and Heyrman, *Commerce and Culture*, 61, 210–213.

16. Morison, *Maritime History*, 14–15; North, *Economic Growth*, 49; and Taylor, *American Colonies*, 176–177.

17. Mark Peterson has noted that by the end of the seventeenth century "the pursuit of overseas trade" was an indispensable part of the Massachusetts economy. See Peterson, *City-State of Boston*, 88; Taylor, *American Colonies*, 176–178, 210–211, 292–293; Heyrman, *Commerce and Culture*, 59–61; and Morison, *Maritime History*, 17–19.

18. The new Massachusetts charter of 1691 required the colony to tolerate all Protestant faiths. See Peterson, *City-State of Boston*, 209; Morison, *Maritime History*, 20, 23–27; Heyrman, *Commerce and Culture*, 8–9; and Taylor, *American Colonies*, 302–303, 340–341.

19. Sellers, *Market Revolution*, 23, 40–41; Taylor, *American Colonies*, 306, 310–311; and North, *Economic Growth*, 42–43.

20. Taylor, *American Colonies*, 422–424; Morison, *Maritime History*, 20, 29–30; Richard H. McKey, Jr., "Elias Hasket Derby: Merchant of Salem, Massachusetts, 1739–1799" (PhD diss., Clark Univ., Worcester, MA, 1951), 6–8, 97, 110, 144, 151; and Robert Booth, *Death of an Empire: The Rise and Murderous Fall of Salem, America's Richest City* (New York: St. Martin's, 2011), xi, xiv–xv.

21. McKey, "Elias Hasket Derby," 2, 147, 149, 155, 158, 160, 175; Morison, *Maritime History*, 44–48; John W. Tyler, "Persistence and Change within the Boston Business Community, 1775–1790," in Conrad Edick Wright and Katheryn P. Viens, eds., *Entrepreneurs: The Boston Business Community*, 100–101, 118–119; and Eric Jay Dolin, *When America First Met China: An Exotic History of Tea, Drugs, and Money in the Age of Sail* (New York: W. W. Norton, 2012), 72–75, 92, 95–97.

22. McKey, "Elias Hasket Derby," 158, 175, 191, 236, 247; Dolin, *When America First Met China*, 15, 57–60, 101–102, 112–113, 161–163; Booth, *Death of an Empire*, 21–22; and Morison, *Maritime History*, 47, 59, 90–91.

23. Dolan, *When America First Met China*, 101–102, 109–110; Morison, *Maritime History*, 46–47; McKey, "Elias Hasket Derby," 263–265, 288; and Allen Johnson and Dumas Malone, eds., *Dictionary of American Biography* (New York: Scribner, 1930), Vol. 5, 249–250.

24. Morison, *Maritime History*, 168–172.

25. Morison, *Maritime History*, 20, 156–159; Dolan, *When America First Met China*, 169–170; and Eric Jay Dolin, *Leviathan: The History of Whaling in America* (New York: W. W. Norton, 2007), 102, 109, 110, 238–239, 341.

26. Donald R. Hickey, *The War of 1812: A Forgotten Conflict* (Urbana: Univ. of Illinois Press, 1989), 6–8.

27. Morison, *Maritime History*, 67, 88, 91, 188–189.

28. Booth, *Death of an Empire*, 8–11; and Hickey, *War of 1812*, 20–21, 197–200, 295–296.

29. Morison, *Maritime History*, 213; and Booth, *Death of an Empire*, 34–35.

30. Hickey, *War of 1812*, 211–212, 296–298; George Dangerfield, *The Awakening of American Nationalism, 1815–1828* (New York: Harper & Row, 1965), viii, 8, 13–15, 32; and North, *Economic Growth*, 102–103, 186.

31. North, *Economic Growth*, 61, 64, 67, 71; and Dangerfield, *Awakening of American Nationalism*, 19–20, 72–75.

32. Dangerfield, *Awakening of American Nationalism*, 73–76; Morison, *Maritime History*, 215–217; and Labaree, "Making of an Empire," 349–352, 358–359.

33. Booth, *Death of an Empire*, 57–60; North, *Economic Growth*, 182–186; and Johnson and Malone, *Dictionary of American Biography*, Vol. 5, 250–251.

2. FACTORIES, CANALS, AND RAILROADS

1. Mark Peterson has called the Bay State's switch from merchant shipping to textile manufacturing its "most radical change" since the Puritan arrival. See Peterson, *City-State of Boston*, 445.

2. Johnson, *Dictionary of American Biography*, Vol. 6, 456–457; and Robert F. Dalzell, Jr., *Enterprising Elites: The Boston Associates and the World They Made* (Cambridge, MA: Harvard UP, 1987), 4–5.

3. Dalzell, *Enterprising Elites*, 8–9; and David Walker Howe, *What Hath God Wrought: The Transformation of America, 1815–1848* (New York: Oxford UP, 2007), 134–135.

4. Dalzell, *Enterprising Elites*, 5; Howe, *What Hath God Wrought*, 132–133; and Kasson, *Civilizing the Machine*, 66–67.

5. Dalzell, *Enterprising Elites*, 15, 20–22; Kasson, *Civilizing the Machine*, 62–64; and Nina Sankovitch, *The Lowells of Massachusetts: An American Family* (New York: St. Martin's Press, 2017), 130–136.

6. Dalzell, *Enterprising Elites*, 25–26, 30; Kasson, *Civilizing the Machine*, 66–67, 71–72, 75; Sven Beckert, *Empire of Cotton: A Global History* (New York: Vintage, 2014), 147, 158; and Anthony J. Connors, *Ingenious Machinists: Two Inventive Lives from the American Industrial Revolution* (Albany: State Univ. of New York Press, 2014), 52, 101.

7. Dalzell, *Enterprising Elites*, 34–35; Matthews, *Toward A New Society*, 142–143; Howe, *What Hath God Wrought*, 132–135; David S. Reynolds, *Waking Giant: America in the Age of Jackson* (New York: Harper Collins, 2008), 62–64; and Jonathon Prude, *The Coming of Industrial Order: Towns and Factory Life in Rural Massachusetts, 1810–1860* (New York: Cambridge UP, 1983), xiii, 51.

8. Dalzell, *Enterprising Elites*, 35; and Prude, *Coming of Industrial Order*, 51.

9. Howe, *What Hath God Wrought*, 134; Dalzell, *Enterprising Elite*, 31–32; and Kasson *Civilizing the Machine*, 67–68.

10. It is ironic that John C. Calhoun supported Lowell's 1816 tariff, when less than two decades later this kind of tariff would bring about the Nullification Crisis. Calhoun would be the main instigator of that crisis. See Peterson, *City-State of Boston*, 445; Kasson, *Civilizing the Machine*, 68; Beckert, *Empire of Cotton*, 159; and Sankovitch, *Lowells of Massachusetts*, 160.

11. Howe, *What Hath God Wrought*, 132–133; North, *Economic Growth*, 169, 172; Kasson, *Civilizing the Machine*, 71–73, 82; and Andrew R. Black, *John Pendleton Kennedy: Early American Novelist, Whig Statesman & Ardent Nationalist* (Baton Rouge: Louisiana State UP, 2016), 127.

12. Andrew Burstein, *America's Jubilee: A Generation Remembers the Revolution after 50 Years of Independence* (New York: Vintage, 2007), 5–7, 11–17, 268; and North, *Economic Growth*, 68–71.

13. Bernstein, *Wedding of the Waters*, 113–114, 178–179, 199, 347–350, 377; Sellers, *Market Revolution*, 41–43; and George Rogers Taylor, *The Transportation Revolution, 1815–1860* (New York: Harpers & Row, 1951), 32–36.

14. "Canallers" were sometimes called "canawlers," reflecting the pronunciation of the region. See Bernstein, *Wedding of the Waters*, 331–332; and Carol Sheriff, *The Artificial River: The Erie Canal and the Paradox of Progress, 1817–1862* (New York: Hill & Wang, 1996), 138–139, 142–147, 150.

15. Taylor, *Transportation Revolution*, 53–55.

16. Taylor, *Transportation Revolution*, 43–44.

17. Taylor, *Transportation Revolution*, 42–43; and Joel Achenbach, *The Grand Idea: George Washington's Potomac and the Race to the West* (New York: Simon & Schuster, 2004), xx.

18. Taylor, *Transportation Revolution*, 154–155; and Kirkland, *Men, Cities, and Transportation*, Vol. 1, 61–66.

19. Taylor, *Transportation Revolution*, 37–38; and Kirkland, *Men, Cities, and Transportation*, Vol. 1, 78–84.

20. Loammi Baldwin recommended a tunnel through the Hoosac Mountain because 220 locks would be needed to lift and lower boats over it. See William B. Meyer, "The Long Agony of the Great Bore" in *Invention and Technology* 1.2 (Fall 1985): 8; Terrence Edward Coyne, "The Hoosac Tunnel" (PhD diss, Clark Univ., 1992), 24; Stephen Salsbury, *The State, the Investor, and the Railroad: The Boston & Albany, 1825–1867* (Cambridge, MA: Harvard UP, 1967), 39, 43–44; and William D. Middleton, "Hoosac Tunnel: Go Big, Go Deep" in *Trains* 68.11 (Nov. 2008): 57–58.

21. George Lowell Austin, *The History of Massachusetts* (Boston: B. B. Russell, 1884), 416; Edward C. Kirkland, "The 'Railroad Scheme' of Massachusetts," in *Journal of Economic History* 5.2 (Nov. 1945): 146; and Dalzell, *Enterprising Elite*, 86.

22. Wolmar, *Great Railroad Revolution*, 5–7, 16–17.

23. America state governments financed 45 percent of railroad construction versus only 7 percent in Prussia. There was no tariff on railroad iron between 1832 and 1843. See Wolmar, *Railroad Revolution*, 18–23; Howe, *What Hath God Wrought*, 564; and Frank Dobbin and Timothy J. Dowd, "How Policy Shapes Competition: Early Railroad Foundings in Massachusetts," in *Administration Science Quarterly* 42.3 (1997): 507517.

24. Wolmar, *Great Railroad Revolution*, 41–44; and John F. Stover, *American Railroads* (Chicago: Public Affairs, 1961), 20–26, 29, 31.

25. In addition to Peter Cooper's contributions, Ross Winans invented the "friction wheel" to allow heavier weights to be pulled on railroad tracks, and Charles Reeder helped perfect the locomotive boiler. See John F. Stover, *History of the Baltimore and Ohio Railroad* (West Lafayette: Univ. of Indiana Press, 1987), 39–40, 48–50; and James D. Dilts, *The Great Road: The Building of the Baltimore & Ohio, the Nation's First Railroad, 1828–1853* (Stanford, CA: Stanford UP, 1993), 48, 70–73, 90–96, 159, 161, 279, 342, 388.

26. Salsbury, *State, Investor, and Railroad*, 46–47, 53, 57–58; Stover, *American Railroads*, 14–15; Thomas H. O'Connor, *Lords of the Loom: The Cotton Whigs and the Coming of the Civil War* (New York: Scribners, 1968), 32; and Arthur M. Johnson and Barry E. Supple, *Boston Capitalists and Western Railroads: A Study in the Nineteenth-Century Railroad Investment Process* (Cambridge, MA: Harvard UP, 1969), 34–39.

27. Taylor, *Transportation Revolution*, 95; Salsbury, *State, Investor, and Railroad*, 61, 69–70; and Wolmar, *Great Railroad Revolution*, 26–27.

28. Salsbury, *State, Investor, and Railroad*, 78–79; and Kirkland, *Men, Cities, and Transportation*, Vol. 1, 115–116.

29. It should be noted that there were two other early railroads in Massachusetts: the Boston & Lowell and the Boston & Providence. The former sounded the death knell for the Middlesex Canal. See Salsbury, *State, Investor, and Railroad*, 32, 81–84, 95, 99, 108; Dalzell, *Enterprising Elite*,

87–88; Johnson and Supple, *Boston Capitalists*, 38–39; and Wolmar, *Great Railroad Revolution*, 35–36.

30. Christian Wolmar has described early railroad companies as a "true ragbag of outfits," highly local in service and often lacking compatible gauges and shared depots as transfer points. See Wolmar, *Great Railroad Revolution*, 49–50; Taylor, *Transportation Revolution*, 85–86; and Salsbury, *State, Investor, and Railroad*, 100–101, 121–121, 179–180.

31. The value of cotton exports increased from $25 million in 1831 to $71 million in 1836. Federal land sales rose from $4 million to $16 million during the same period. See Salsbury, *State, Investor, and Railroad*, 141–142; North, *Economic Development*, 194–197; and Peter Termin, *Jacksonian Economy* (New York: W. W. Norton, 1969), 124.

32. Salsbury, *State, Investor, and Railroad*, 163, 169, 173–174, 176.

33. Daniel Webster had established his reputation early with his reply to Senator Robert Hayne of South Carolina, in a rousing defense of the Union, during the 1830 Webster-Hayne Debate. Webster was one of the most imposing figures in American political life. Ron Formisano has gone so far as to call the Massachusetts Whigs "the Puritan Party." See Formisano, *Transformation of Political Culture*, 205, 247, 255, 259, 270, 283; Howe, *What Hath God Wrought*, 357–362; Sean Wilentz, *The Rise of American Democracy: Jefferson to Lincoln* (New York: W. W. Norton, 2005), 349, 437; and William F. Hartford, *Money, Morals, and Politics: Massachusetts in the Age of the Boston Associates* (Boston: Northeastern UP, 2001), 79–80, 100, 103.

34. Economic historians have blamed the Panic of 1837 on the Jackson administration, land speculators, cotton growers, and British bankers. In Massachusetts, Jackson's pet bank was the Commonwealth Bank. It went bankrupt in January 1838, much to the disgrace of its Democratic investors and management. Railroad historians agree that the Western Railroad's debt burden committed it to high freight rates from the start. See O'Connor, *Lords of the Loom*, 40; Termin, *Jacksonian Economy*, 13–17, 124; North, *Economic Development*, 68, 70–75, 194–199, 202–203; Wilentz, *Rise of American Democracy*, 349, 437; Salsbury, *State, Investor, and Railroad*, 143–156; Kirkland, *Men, Cities, and Transportation*, Vol. 1, 136–137; Stover, *American Railroads*, 16–17; Jessica M. Lepler, *The Many Panics of 1837: People, Politics, and the Creation of a Transatlantic Financial Crisis* (New York: Cambridge UP, 2013), 47, 53, 107–108, 178, 189; and Dobbin and Dowd, "How Policy Shapes Competition," 507.

35. Salsbury, *State, Investor, and Railroad*, 183, 186, 189–190, 194–195, 251, 254, 279, 291, 299; Johnson and Supple, *Boston's Capitalists*, 43–44; Howe, *What Hath God Wrought*, 566–567; Taylor, *Transportation Revolution*, 85, 134; and Hill, *Boston's Trade and Commerce for Forty Years*, 10–11.

36. Salsbury, *State, Investor, and Railroad*, 94, 158–160, 209, 214, 223–225, 255–257; Kirkland, *Men, Cities, and Transportation*, Vol. 1, 140–143; Formisano, *Transformation of Political Culture*, 127–128; and Johnson and Malone, *Dictionary of American Biography*, Vol. 5, 252–253.

37. Salsbury, *State, Investor, and Railroad*, 176, 195, 208–209, 231; and Johnson and Malone, *Dictionary of American Biography*,

38. Nathaniel Hawthorne was skeptical of technical advancement. In 1843, he had written a satire entitled "The Celestial Railroad." See Taylor, *Transportation Revolution*, 84; Marx, *Machine in the Garden*, 15, 229, 249–250; Kasson, *Civilizing the Machine*, 49, 114–115; Nathaniel Hawthorne, *The House of Seven Gables* (1851; rpr. New York: Bantam, 1981), 196–201.

39. It should be noted that Bancroft was never a Whig but a Democrat, serving in Congress and in the Polk administration. See Marx, *Machine in the Garden*, 170–171, 181–189; and Timothy Walker, "Defense of Mechanical Philosophy," in *North American Review* 33.72 (July 1831): 122–136.

40. The Parliamentary Commission's report, entitled *Machinery of the United States*, is quoted at length by Douglass North. See North, *Economic Growth*, 172–174.

41. American National Biography Online @www.anb.org.

3. MASSACHUSETTS LOOKS WEST

1. Wheelwright, *Alvah Crocker*, 3–4; and Nicholas A. Basbanes, *On Paper: The Everything of Its Two-Thousand-Year History* (New York: Vintage, 2014), 87, 305.

2. Wheelwright, *Alvah Crocker*, 8–14, 33–35.

3. John Stover has pointed out that the Fitchburg Railroad cost $30,000 per mile to build versus over $50,000 for the Western Railroad. See Stover, *American Railroads*, 29; Wheelwright, *Alvah Crocker*, 25–27; Salsbury, *State, Investor, and Railroad*, 269–270, 280; Middleton, "Hoosac Tunnel," 62–63; and bing.com/images/alvahcrocker.

4. Johnson and Malone, *Dictionary of American Biography*, 250; Wheelwright, *Alvah Crocker*, 46–47; North, *Economic Development*, 206; Elias Hasket Derby, *Two Months Abroad: Or, A Trip to England, France, Baden, Prussia, and Belgium in August and September, 1843* (1843; rpr. Columbia, SC: Wentworth, 2016), 51–52; and Cliff Schexnayder, *Builders of the Hoosac Tunnel* (Portsmouth, NH: Peter E. Randall, 2015), 109–110.

5. Derby, *Two Months Abroad*, 44–45, 48, 51–53.

6. Edward Kirkland has argued that northern Massachusetts was "a region intensely conscious of its need for a railroad." See Kirkland, "The Hoosac Tunnel: The Great Bore," in *New England Quarterly* 20 (March–Dec. 1947): 87–91; Wheelwright, *Alvah Crocker*, 38–40; Coyne, "Hoosac Tunnel," 38–39; and James A. Ward, *Railroads and the Character of America, 1820–1887* (Knoxville: Univ. of Tennessee Press, 1986), 133, 146, 152.

7. Railroad mileage in the states of Michigan, Indiana, Illinois, Wisconsin, and Iowa increased from 374 to 1,276 between 1845 and 1850. See North, *Economic Growth*, 74, 142, 193–194, 211–212, 251; and Salsbury, *State, Investor, and Railroad*, 278.

8. Edward Hitchcock's geological analysis of the Hoosac Tunnel was reprinted in legislative documents in 1859 and 1863. It is also summarized in William Browne's history of the tunnel. Browne was a local historian in North Adams, Massachusetts. His history was presented to the town's chamber of commerce in 1924 and later published by the town's library. See *Argument on the Petition of the Troy & Greenfield Railroad Company for a Change in the Conditions of the Loan Act before the Joint Committee of Railways and Canals, January 12, 1859* (Boston: Alfred Mudge & Son, 1859), 16–17; *Report of the Commissioners upon the Troy and Greenfield Railroad and Hoosac Tunnel, February 28, 1863* (Boston: Wright & Potter, 1863), 39–40; Salsbury, *State, Investor, and Railroad*, 280–281; Coyne, "Hoosac Tunnel," 32–35, 44–45; and William B. Browne, "Chamber of Commerce Speech, March 4, 1924" (North Adams, MA: publisher unknown, 1924), 4–6. See also *Senate . . . No. 120. Minority Report.* (Boston: publisher unknown, 1848), 37–40, 45–46; *House . . .*

212 (Boston: publisher unknown, 1848), 14–17, 25; and *An Act to Incorporate the Troy and Greenfield Railroad Company* (Boston: Eastburn's Press, 1848).

9. Kirkland, *Men, Cities, and Transportation,* Vol. I, 388, 391–396.

10. *Boston Daily Advertiser,* April 22, 24, 29, 1851; and *The Hoosac Mountain Tunnel Speech of the Hon. E. D. Beach of the Senate, on the Bill Providing for a Loan of Credit of Massachusetts to the Amount of Two Millions of Dollars to the Troy & Greenfield Railroad in the Senate, April 11, 12, 1851* (Boston: Eastburn's Press, 1851), 5, 8, 16.

11. *Hoosac Mountain Tunnel Speech of the Hon. E. D. Beach,* 18, 21; and *The Memorial of the Western Railroad Corporation relating to the Application of the Troy and Greenfield Railroad for a State Loan of Two Millions of Dollars* (Boston: John Wilson & Son, 1851), 7–8, 10–12.

12. *Memorial of the Western Railroad,* 22, 24; and *Truths about the Hoosac Tunnel Project* (Boston: Eastburn's Press, 1851), 6, 9–10, 15.

13. *Speech of Hon. Whiting Griswold in the Senate of Massachusetts, April 9 and 11, 1851* (Boston: Eastburn's Press, 1851), 4, 7, 10–11; Smith, *Virgin Land,* 186; Marx, *Machine in the Garden,* 225; and Kirkland, "Hoosac Tunnel," 96–97.

14. *Truths about the Hoosac Tunnel,* 7, 9, 11; and Kirkland, "Hoosac Tunnel," 96–97.

15. William Hartford has noted that the spindleage of Massachusetts textile mills increased by 57 percent between 1845 and 1850, from 817,483 to 1,288,091 spindles, resulting in overcapacity. See Hartford, *Money, Morals, and Politics,* 158, 173–174; O'Conner, *Lords of the Loom,* 32; and Coyne, "Hoosac Tunnel," 64–67, 74, 88–89, 135.

16. The Fugitive Slave Law required the return of escaped slaves living in the North to their southern masters. The arrest of slaves in Boston caused riots and wide-scale civil disobedience. Whig presidential candidate Winfield Scott carried four states in the election of 1852, garnering only 42 electoral votes versus 254 for Democrat Franklin Pierce. See Formisano, *Political Culture,* 297, 327–328, 330; O'Connor, *Lords of the Loom,* 44–45, 57, 65–66, 80–81, 88, 96; Hartford, *Money, Morals, and Politics,* 114–115, 120, 130, 149–150; and Michael Holt, *The Rise and Fall of the American Whig Party: Jacksonian Politics and the Onset of the Civil War* (New York: Oxford UP, 1999), 753–754.

17. *The Hoosac Tunnel: A Brief Report of the Hearing of the Troy & Greenfield Railroad Company Petitioners for a Loan of Two Millions before a Joint Committee of the Legislature of Massachusetts* (Boston: Thurston, Torry, & Emerson, 1853), 4–7, 18, 24–25; *Hoosac Tunnel. The Memorial of the Western Railroad Corporation relating to the Application of the Troy and Greenfield Railroad for a State Loan of Two Millions of Dollars* (Boston: Eastburn's Press, 1853), 6–8, 11, 16, 18; and Wheelwright, *Alvah Crocker,* 46–48.

18. *Hoosac Tunnel: A Brief Report,* 8–10; Browne, "Chamber of Commerce Speech," 7–8; Gary S. Brierley, "Construction of the Hoosac Tunnel 1855 to 1876" in *Journal of the Boston Society of Civil Engineers Section American Society of Civil Engineers* 63.3 (Oct. 1976): 184; and Carl R. Byron, *A Pinprick of Light: The Troy and Greenfield Railroad and Its Hoosac Tunnel* (Shelburne, VT: New England Press, 1955), 8–9.

19. *Hoosac Tunnel: A Brief Report,* 10–11, 27–29, 37.

20. *Hoosac Tunnel: A Brief Report,* 72–73.

21. *Hoosac Tunnel: A Brief Report,* 12–19, 24–26, 28, 48–49.

22. *Hoosac Tunnel: A Brief Report*, 34–43, 53–54.

23. North, *Economic Growth*, 206–207; Howe, *What Hath God Wrought*, 695–698, 809, 823–827; and Robert W. Merry, *A Country of Vast Design: James K. Polk, The Mexican War, and the Conquest of the American Continent* (New York: Simon & Schuster, 2009), 462–463, 477.

24. *Hoosac Tunnel: A Brief Report*, 52; and North, *Economic Growth*, 177–179, 204–208, 211–212.

25. *Hoosac Tunnel: A Brief Report*, 56.

26. *Hoosac Tunnel: A Brief Report*, 67; *Hoosac Tunnel. Speech of Ansel Phelps, Jr. on Petition of the Troy and Greenfield Railroad Corporation for State Aid, April 6, 1853* (Boston: Thurston, Torry, & Emerson, 1853), 14–15, 29–30, 33; and *Speech on Bill Granting State Aid for the Construction of the Hoosac Tunnel by A. A. Richmond, Esq., of Adams* (Boston: Eastburn's Press, 1854), 10, 13.

27. Browne, "Chamber of Commerce Speech," 8–9; Coyne, "Hoosac Tunnel," 72–77; and Schexnayder, *Builders of the Hoosac Tunnel*, 134–135.

28. Coyne, "Hoosac Tunnel," 77, 80–81; Middleton, "Hoosac Tunnel," 65–66; Schexnayder, *Builders of the Hoosac Tunnel*, 137; and Byron, *Pinprick of Light*, 12.

29. Byron, *Pinprick of Light*, 9–13.

30. The contract with Edward Serrell was signed in January 1855 but would be superseded by that with Herman Haupt a year and a half later. At that point, Serrell and Haupt became business partners. See Byron, *Pinprick of Light*, 15–16; Schexnayder, *Builders of the Hoosac Tunnel*, 140–141; and Coyne, "Hoosac Tunnel," 94–95.

31. *Speech of Hon. H. G. Parker of Greenfield* (Boston: publisher unknown, 1854), 7; and Coyne, "Hoosac Tunnel," 59.

4. HERMAN HAUPT AGAINST THE MOUNTAIN

1. Lord, *Lincoln's Railroad Man*, 21–25; and James A Ward, *That Man Haupt: A Biography of Herman Haupt* (Baton Rouge: Louisiana State Univ. Press, 1973), 5, 8, 11, 18–19, 49.

2. Herman Haupt, *Reminiscences of General Herman Haupt* (Milwaukee, WI: Wright & Joy, 1901), xiii–xx.

3. Haupt, *Reminiscences*, xx; Schexnayder, *Builders of the Hoosac Tunnel*, 170, 172–173; Coyne, "Hoosac Tunnel," 95–96; and Ward, *That Man Haupt*, 52–57.

4. Coyne, "Hoosac Tunnel," 94–95; and Schexnayder, *Builders of the Hoosac Tunnel*, 173.

5. The "Tunnel Affray" occurred when Serrell tried to fire a contractor named Hill at the east portal and replace him with the Stanton brothers. Hill's Irish workers refused to leave the site when the Stantons' crew arrived. A terrible fight ensued. See Coyne, "Hoosac Tunnel," 88–89, 92–93, 97–98; Ward, *That Man Haupt*, 61–62; and Schexnayder, *Builders of the Hoosac Tunnel*, 141–142, 173–174.

6. Schexnayder, *Builders of the Hoosac Tunnel*, 174, 176–178, 180–181; and Coyne, "Hoosac Tunnel," 98–101.

7. Cliff Schexnayder has correctly characterized Serrell's separation package as too generous and faulted Haupt for it. However, it is possible that the Troy & Greenfield Railroad wanted to

ensure Serrell remained silent regarding construction problems at the Hoosac Mountain and the company's financial difficulties. See Schexnayder, *Builders of the Hoosac Tunnel*, 178–181; and Coyne, "Hoosac Tunnel," 99, 101–102.

8. Byron, *Pinprick of Light*, 24; Coyne, "Hoosac Tunnel," 105; Haupt, *Reminiscences*, xxi; Schexnayder, *Builders of the Hoosac Tunnel*, 187; and Ward, *That Man Haupt*, 50, 62.

9. *Francis William Bird: A Biographical Sketch* (Boston: Norwood, 1897), 6–12, 16, 18–20, 25, 32, 39; and John A. Garraty and Mark C. Carnes, eds., *American National Biography* New York: Oxford UP, 1999), Vol. 2, 805–806.

10. John R. Mulkern, *The Know Nothing Party in Massachusetts: The Rise and Fall of a People's Movement* (Boston: Northeastern UP, 1990), 55; Formisano, *Transformation of Political Culture*, 332; and the *Springfield Republican*, June 2, 1855.

11. Historians have struggled to characterize the Know Nothings. For Michael Holt and Bruce Levine, they were nativists. For Ron Formisano and John Mulkern, they were populist reformers. Tyler Anbinder believes they were strongly antislavery. See Holt, *American Whig Party*, 805–806, 845–848; Bruce Levine, "Conservatism, Nativism, and Slavery: Thomas R. Whitney and the Origins of the Know-Nothing Party," in *Journal of American History* 88.2 (Sept. 2001): 457–458, 467; Formisano, *Transformation of Political Culture*, 340–343; Mulkern, *Know-Nothing Party of Massachusetts*, 69, 111; Tyler Anbinder, *Nativism & Slavery: The Northern Know Nothings & the Politics of the 1850s* (New York: Oxford UP, 1992), ix, 23, 89–102, 127–128.

12. Historian Richard Hofstadter observed that "anti-Catholicism has always been the pornography of the Puritan." See Anbinder, *Nativism & Slavery*, 11, 27, 45–46, 89, 99, 107–108, 115; Formisano, *Transformation of Political Culture*, 331–335; John F. Berns, *Providence & Patriotism in Early America, 1640–1815* (Charlottesville: Univ. of Virginia Press, 1978), 16–17, 47–48; and Alfred F. Young, *The Shoemaker and the Tea Party: Memory and the American Revolution* (Boston: Beacon Press, 1999), 21–22.

13. Roger Daniels, *Coming to America: A History of Immigration in American Life* (New York: Harper Collins, 1990), 128, 137; and Oscar Handlin, *Boston Immigration* (Cambridge, MA: Belknap Press, 1979), 48–49, 55, 71, 73–77.

14. James Scott argues that nineteenth-century visionaries saw the need to reengineer nature by replacing curves with straight lines and leveling hills to create flat spaces. See James C. Scott, *Seeing Like A State: How Certain Schemes to Improve the Human Condition Have Failed* (New Haven: Yale UP, 1998), 94–97; *Hoosac Tunnel: Arguments of E. Hasket Derby, Esq., Delivered Feb. 29th, 1856 Before A Joint Special Committee of the Legislature of Massachusetts, in behalf of the Troy & Greenfield Railroad Company; Petitioners for a State Subscription to Their Stock* (Boston: Bazin & Chandler, 1856), 3–6; and Coyne, "Hoosac Tunnel," 99–100.

15. Formisano, *Transformation of Political Culture*, 335; *Boston Daily Advertiser,* May 16, 1856; Coyne, "Hoosac Tunnel," 83–84; and Fred H. Harrington, *Fighting Politician: Major General N. P. Banks* (Philadelphia: Univ. of Pennsylvania Press, 1948), 17–18.

16. Coyne, "Hoosac Tunnel," 104–106; Schexnayder, *Builders of the Hoosac Tunnel*, 186–190; and *To the Taxpayers of Massachusetts. Read the Reasons of Gov. Gardner for Vetoing the Bill appropriating $2,000,000 to the Troy & Greenfield Railroad* (Boston: publisher unknown, 1857), 5–7.

17. Antislavery sentiment spiked in Massachusetts after "Bloody Kansas," where in May 1856

proslavery forces sacked the town of Lawrence, Kansas, the state's free-labor headquarters, and that same month Massachusetts senator Charles Sumner was nearly caned to death by a southern sympathizer on the floor of Congress. In the presidential election of 1856, Millard Fillmore won only Maryland. See Mulkern, *Know Nothing Party of Massachusetts*, 140–1144; David S. Potter, *The Impending Crisis, 1848–1861* (New York: Harpers & Row, 1976), 254–255; William E. Gienapp, *The Origins of the Republican Party, 1852–1856* (New York: Oxford UP, 1987), 419–420; and *Boston Daily Advertiser*, March 2, 3, May 22, 1857.

18. The Panic of 1857 began with the collapse of the Ohio Life and Insurance and Trust Company. The panic dealt the death blow to Samuel Lane's North West Railroad. Other railroads in the Midwest struggled to survive. Boston financiers John E. Thayer and John M. Forbes were heavily invested in these roads and made it through only after loans from Barings Bank in England. See North, *Economic Growth*, 212–213; Johnson and Supple, *Boston Capitalists*, 79–80, 176; Byron, *Pinprick of Light*, 18; Ward, *That Man Haupt*, 77–78; Coyne, "Hoosac Tunnel," 109; and Harrington, *Fighting Politician*, 33–34.

19. North, *Economic Growth*, 213; Schexnayder, *Builders of the Hoosac Tunnel*, 198; Coyne, "Hoosac Tunnel," 110–111; *Hoosac Valley News*, March 17, 1858; and Kirkland, *Men, Cities, and Transportation*, Vol. 1, 326.

20. In the end, Haupt probably netted no more than $90,000 in paid-in subscriptions of the $145,000 pledged by the townships. Still, voting for the subscriptions showed the extent of support for the tunnel in these communities. For example, North Adams voted 582 to 187 for it. See Ward, *That Man Haupt*, 80; Coyne, "Hoosac Tunnel," 114–116; and Browne, "Chamber of Commerce Speech," 9–10.

21. Coyne, "Hoosac Tunnel," 121–122; Ward, *That Man Haupt*, 79–83; and *Hoosac Valley News*, April 7, 1858.

22. Anbinder, *Nativism & Slavery*, 155, 210–211, 250–251; Mulkern, *Know Nothing Party of Massachusetts*, 169; and Dale Baum, *The Civil War Party System: The Case of Massachusetts, 1848–1876* (Chapel Hill: Univ. of North Carolina Press, 1984), 40–42, 45.

23. Ward, *That Man Haupt*, 84–85; Coyne, "Hoosac Tunnel," 125–129; and Schexnayder, *Builders of the Hoosac Tunnel*, 205–206.

24. Schexnayder, *Builders of the Hoosac Tunnel*, 207, 214–215; Ward, *That Man Haupt*, 87- 92; and Coyne, "Hoosac Tunnel," 134, 137–139.

25. Known at first as "Haupt's shaft" and later as the "west shaft," it was 2,500 feet east of the west portal and was operational by September 1859. See Ward, *That Man Haupt*, 87–88; Coyne, "Hoosac Tunnel," 108–109, 124, 139; Schexnayder, *Builders of the Hoosac Tunnel*, 208–209; and Browne, "Speech to the Chamber of Commerce," 11.

26. The description of life around the west portal is from Isaac Browne. The newspaper reports and vital records of mining accidents have been collected by Charles Cahoon, president of the North Adams Historical Society, and are among the holdings of the North Adams Public Library. See Browne, *Hoosac Tunnel Days*, 1–3; *North Adams Weekly Transcript*, Jan. 22 and July 23, 1857; *Hoosac Valley News*, Nov. 25, 1858; and Vital Records of Adams and those of Florida, Massachusetts.

27. Again, casualties at the tunnel have been researched by Charles Cahoon and can be found

at the North Adams Library. Cliff Schexnayder provides the best description of tunneling methods used at the Hoosac Mountain. See Schexnayder, *Builders of the Hoosac Tunnel*, 184–185; Byron, *Pinprick of Light*, 10–11; *Hoosac Valley News*, Dec. 22, 1859, March 29, April 5, 1860; *Hoosac Valley News and Transcript*, April 18, 1861; and Vital Records of Adams and Florida, Massachusetts.

28. Haupt actually inherited the second Wilson drilling machine from Edward Serrell. Still, Haupt must have had some faith in it because he spent nine months trying to perfect it. Edward Kirkland argued that Haupt was "hypnotized" by big boring contraptions. See Kirkland, "Hoosac Tunnel," 99–100; Ward, *That Man Haupt*, 88–89, 96–97, 101, 154; Schexnayder, *Builders of the Hoosac Tunnel*, 191–192, 204–205; and Brierley, "Construction of the Hoosac Tunnel," 184–185.

29. Ward, *That Man Haupt*, 89, 94–96, 99–100; Schexnayder, *Builders of the Hoosac Tunnel*, 208, 210–213, 215–217; and Wolmar, *Railroad Revolution*, 65–66.

30. Coyne, "Hoosac Tunnel," 119, 125, 130–131, 139–140; and *Hoosac Valley News*, July 8, 1858, Aug. 9, 1860.

31. Banks balked at the creation of a black militia, and he favored erecting a statue of Webster on the State House grounds. Radical antislavery elements had excoriated Webster for his Seventh of March Speech (March 7, 1850), wherein he defended the Compromise of 1850 and its fugitive slave law. John Brown's raid on Harpers Ferry, Virginia (October 18, 1859), was condemned by Banks and many other conservative Republicans as an act of lawlessness. However, antislavery radicals hailed it as a noble effort to liberate Virginia's slaves. Several elite Bostonians had secretly provided Brown with funding for his raid. The two-year immigration amendment would embarrass Bay State Republicans at the party's national convention in Chicago. See Harrington, *Fighting Politician*, 47–48; Anbinder, *Nativism & Slavery*, 251–252; Potter, *Impending Crisis*, 132–133, 364, 378–382; Eric Foner, *The Fiery Trial: Abraham Lincoln and American Slavery* (New York: W. W. Norton, 2010), 140; and Richard Abbott, "Massachusetts: Maintenance and Hegemony," in James C. Mohr, ed., *Radical Republicans in the North: State Politics during Reconstruction* (Baltimore: Johns Hopkins UP, 1976), 3–4.

32. Ward, *That Man Haupt*, 97–98; and Coyne, "Hoosac Tunnel," 146–149.

33. Ward, *That Man Haupt*, 99–103; Coyne, "Hoosac Tunnel," 148–151.

34. *Hoosac Valley News and Transcript*, July 25 and Oct. 17, 1861; and Baum, *Civil War Party System*, 65–67.

35. Ward, *That Man Haupt*, 103–108; Coyne, "Hoosac Tunnel," 151–160; *Speech of Hon. Charles G. Loring on the Troy and Greenfield Railroad Bill in the Massachusetts Senate, Wednesday, April 16, 1862* (Boston: publisher unknown, 1862), 4, 6, 15; and *Speech of Hon. Alvah Crocker on the Bill for More Speedy Completion of the Troy and Greenfield Railroad in the Senate of Massachusetts, April 15, 1862* (Boston: Wright & Potter, 1862), 15, 18.

36. F. W. Bird, *The Road to Ruin, or, The Decline and Fall of the Hoosac Tunnel* (Boston: Wright & Potter, 1862), 12, 22, 31–32.

37. Bird, *Road to Ruin*, 12–16, 24–25.

38. For Charles Dickens, Fagin was the archetypical Jewish villain. Arriving in America during 1839, *Oliver Twist* was popular and contained hundreds of references to Fagin's Jewishness, which Bird was clearly aware of. Haupt's German name and ancestry fit well with Bird's purpose of making his opponent not just an outsider to Massachusetts but a supremely evil one at that. See Bird,

Road to Ruin, 36–39; and Margaret Drabble, *Oxford Companion to English Literature* (New York: Oxford UP, 1985), 272–273.

39. One of the great ironies of Haupt's story was his eventual reconciliation with Francis Bird in early 1867. Both Haupt and Bird hated the state's head commissioner of the Hoosac Tunnel, John Brooks. At their first meeting, Bird said he was sorry for his treatment of Haupt and that Daniel Harris of the Western Railroad had supplied most of the information for the *Road to Ruin* pamphlet. Haupt and Bird remained close friends until the latter's death. See *Address of His Excellency John A. Andrew to the Legislature of Massachusetts (January 3, 1861)* (Boston: Wright & Potter, 1862), 29; Ward, *That Man Haupt*, 108–109; and Coyne, "Hoosac Tunnel," 181–183.

40. James McPherson has called Herman Haupt the Civil War's "wizard of railroading." See James M. McPherson, *Battle Cry of Freedom: The Civil War Era* (New York: Oxford UP, 1988), 527, 532; and Ward, *That Man Haupt*, 117, 124, 161–163

41. Ward, *That Man Haupt*, 150–152, 166, 170; and Haupt, *Reminiscences*, 261–301.

42. After numerous attempts to gain compensation from Massachusetts, Herman Haupt reached a settlement with the state in 1868. He received $53,000, a seventh of what he thought he was owed. In 1884, Haupt received a similar sum for his shares of H. Haupt & Company, which the state felt compelled to purchase to secure its ownership of the tunnel. However, these payments were not enough to cover Haupt's accumulated debts or free his encumbered property. See Ward, *That Man Haupt*, 182, 186–187, 190–193, 197, 203–211, 245–247; and Haupt, *Reminiscences*, 44–45.

43. Haupt's biographer James Ward has correctly argued that American business changed after the Civil War. Business enterprises transitioned from small, lightly capitalized ventures with a handful of partners, in which technically brilliant men like Haupt could do well, to "mammoth collective efforts," with massive capitalization, powerful political influence, and a breadth of specialized management talent. Sadly, Haupt had difficulty finding his way in this new business order. See Ward, *That Man Haupt*, 248–249; and Glenn Porter, *The Rise of Big Business, 1860–1920* (Wheeling, IL: Davidson, Harlan, 1973), 8, 19–20, 98–99, 116–117.

5. THE COMMONWEALTH INTERVENES

1. In spite of their often provocative and incendiary tactics, recent historiography has correctly praised the role of the abolitionists. For example, Eric Foner has argued that the abolitionists influenced the Republican Party to not only prevent slavery from expanding to the territories but to eliminate it altogether. For an earlier, more negative view of the abolitionists, Gilbert Hobbs Barnes is one of the most readable. The irony in Emerson's description of the Bay State's intellectual ferment is that he went a long way to encourage it with his ideas about self-culture and individualism. He later turned more conservative. See Foner, *Fiery Trail*, xvii, 19–25, 84–85; Gilbert Hobbs Barnes, *The Anti-Slavery Impulse, 1830–1844* (New York: Harcourt Brace, 1933), 88, 90–94, 125–129, 174–175, 196–197; George M. Fredrickson, *The Inner Civil War: Northern Intellectuals and the Crisis of the Union* (New York: Harper & Row, 1965), 10–11, 17–19, 114–119, 147, 151–152; 176–177; and Ralph Waldo Emerson, *The Complete Works of Ralph Waldo Emerson:*

Natural History of Intellect and Other Papers, Vol. 12 (Ann Arbor: Univ. of Michigan Press, 2006), 195, 203, 206, 208.

2. Wheelwright, *Life and Times of Alvah Crocker,* 39, 49; Coyne, "Hoosac Tunnel," 221–222; and the *Hoosac Valley News and Transcript,* Oct. 19, 1865.

3. Schexnayder, *Builders of the Hoosac Tunnel,* 233–234, 237; Johnson and Supple, *Boston Capitalists,* 90, 108, 175; Coyne, "Hoosac Tunnel," 179–180; Wolmar, *Railroad Revolution,* 70, 121–122; Stover, *History of Baltimore and Ohio,* 39–40, 48–50; and Richard C. Overton, *Burlington Route: A History of the Burlington Lines* (Lincoln: University of Nebraska Press, 1965), 28, 80, 85, 94.

4. The State of Massachusetts had spent $954,963, plus interest, on the Hoosac Tunnel in the twelve years since groundbreaking. See *Report of the Commissioners upon the Troy and Greenfield Railroad and Hoosac Tunnel, February 28, 1863* (Boston: Wright & Potter, 1863), 3–4, 49–52, 61, 105, 105, 128; Coyne, "Hoosac Tunnel," 187, 190–194; Byron, *Pinprick of Light,* 30–31; and Schexnayder, *Builders of the Hoosac Tunnel,* 263–265, 273.

5. *Report of the Commissioners . . . February 28, 1863,* 135–136, 180, 185, 187; Schexnayder, *Builders of the Hoosac Tunnel,* 254–255; and Coyne, "Hoosac Tunnel," 194–198.

6. Richard Bensel has argued that the Civil War was a major impetus for state expansion. According to Bensel, congressional approval of the transcontinental railroad was "an active assertion of central state authority." The transcontinental railroad was approved in two legislative acts, the first in 1862 and a second in 1864. The Central Pacific Railroad built east from Sacramento, California, and the Union Pacific west from Council Bluffs, Iowa. The land grants were larger than the states of New Jersey and New Hampshire combined. See Richard Franklin Bensel, *Yankee Leviathan: The Origins of Central State Authority in America, 1859–1877* (New York: Cambridge UP, 1990), 3, 178, 251; Wolmar, *Railroad Revolution,* 128–130; Johnson and Supple, *Boston Capitalists,* 196–197; and Richard White, *Railroaded: The Transcontinentals and the Making of Modern America* (New York: W. W. Norton, 2011), 17–24, 27.

7. Stover, *American Railroads,* 37–41; and Wolmar, *Railroad Revolution,* 85–87, 12–122.

8. Schexnayder, *Builders of the Hoosac Tunnel,* 260–263, 496.

9. Post–Civil War inflation had raised labor costs along with everything else. See *Report of Messrs. John W. Brooks, Samuel M. Felton, and Alexander Holmes, Commissioners appointed under chapter one hundred and fifty-six of Acts of 1862* (Boston: Wright & Potter, 1865), 29–30; Schexnayder, *Builders of the Hoosac Tunnel,* 288; and Coyne, "Hoosac Tunnel," 204–206.

10. The best information about injuries to Hoosac Tunnel workers has been compiled by Charles Cahoon from newspaper accounts of accidents and fatalities at the tunnel and vital records in North Adams, Florida, and other nearby towns. The report is entitled "Hoosac Tunnel Accident Victims" and is in the possession of the North Adams Public Library. It covers the period from 1857 through 1875. See Charles Cahoon, "Hoosac Tunnel Accident Victims" (unpublished report).

11. Cahoon, "Hoosac Tunnel Accident Victims"; and Schexnayder, *Builders of the Hoosac Tunnel,* 267–268, 308–309, 467.

12. For a visual comparison, modern readers should think of the Empire State Building at a height of 1,250 feet. Charles Storrow anticipated that the central shaft would take three years to dig. It took twice that long. See *Report of the Commissioners . . . February 28, 1863,* 117–119; and *Adams News and Transcript,* Sept. 7, 1865.

13. *Report of the Commissioners . . . February 28, 1863*, 118; *Report of the Auditor of Accounts of the Commonwealth of Massachusetts for the Year Ending December 31, 1864* (Boston: publisher unknown, 1864), 104–105; and Schexnayder, *Builders of the Hoosac Tunnel*, 279–282.

14. The east peak of the Hoosac Mountain was known as Whitcomb Hill and the west peak Spruce Hill. The hill east of the east portal was Roe Mountain, and the one west of the west portal was Notch Road. Only the Notch Road aligning tower remains identifiable by its stone walls today. Cliff Schexnayder has explained that Doane's high-powered transit scopes came from John H. Temple of Boston. See *Report of Messrs. John W. Brooks*, 23, 29, 31; Schexnayder, *Builders of the Hoosac Tunnel*, 266–267; Byron, *Pinprick of Light*, 37; and IMR Films documentary, *On to the Hoosac, On to the West*, Part 3, viewable at www.hoosactunneldocumentary.com.

15. *Report of Messrs. John W. Brooks*, 18.

16. *Report of Messrs. John W. Brooks*, 14; Ward, *That Man Haupt*, 88, 97, 101; Brierley, "Construction of the Hoosac Tunnel," 185–186; "Commissioners' Report" in *House—No. 3* (Boston, publisher unknown, 1865), 3, 6, 8; and Schexnayder, *Builders of the Hoosac Tunnel*, 205, 273, 283, 297.

17. Cliff Schexnayder has argued that Alvah Crocker was the prime mover behind the development of the Burleigh drill. James Ward and Terrence Coyne have suggested that John Brooks played this role. See Schexnayder, *Builders of the Hoosac Tunnel*, 296–297; Ward, *That Man Haupt*, 173–174, 177; and Coyne, "Hoosac Tunnel," 215, 217.

18. Haupt also patented his drill design in England. He even won a gold medal for it at the 1867 Paris Exhibition, presented to him personally by Napoleon III. In spite of this honor, Haupt never made any money from his drill design either in America or Europe. His drill test at the Wiconisco coal mine cost him $14,000 that he did not have. Keeping his son Jacob in business was also costly and paid few dividends. See Ward, *That Man Haupt*, 88–89, 122, 154, 173–178, 181.

19. Burleigh purchased the remaining years on the Fowles patent in August 1867 for $10,000. The transmission of compressed air was a success at the tunnel. There were four twenty-horse-power turbines just below the Deerfield Dam in the turbine building. They drove sixteen air compressors. Compressed air went to the Burleigh drills at sixty-five pounds per square inch. There were only two pounds of lost pressure per mile and a half of rubber pipe leading to the drills. See *Report on the Hoosac Tunnel and Troy and Greenfield Railroad by the Joint Standing Committee of 1866* (Boston: Wright & Potter, 1867), 34–38; Coyne, "Hoosac Tunnel," 217, 177, 235–239; Schexnayder, *Builders of the Hoosac Tunnel*, 302–303, 310–315, 395; and *Scientific American* 22, no. 7 (Feb. 12, 1870): 106.

20. John Brooks recovered his health in 1867 and became president of the Chicago, Burlington & Quincy Railroad. In 1875, however, John Forbes caught Brooks and a partner stealing funds from the railroad and fired both men. Brooks never recovered financially or health-wise, dying in 1881 while vacationing in Europe. Work on the tunnel had burned through the original state loan of $2 million, plus an additional $800,000 appropriated in 1865. With the new appropriation of $900,000, the total spending now stood at $3.7 million. See *Act of May 14, 1866, providing $900,000 to completion of Troy & Greenfield Railroad* (Boston: Wright & Potter, 1866), 14; Byron, *Pinprick of Light*, 39; and Schenayder, *Builders of the Hoosac Tunnel*, 291, 308–309, 498.

21. F. W. Bird, *The Hoosac Tunnel: Our Financial Maelstrom* (Boston: E. P. Dutton, 1866), 11, 30, 34, 39, 59, 65.

22. Bird, *Hoosac Tunnel: Financial Maelstrom,* 65; Schexnayder, *Builders of the Hoosac Tunnel,* 289–290; and *Act of May 14, 1866, providing $900,000,* 19.

23. Baum, *Civil War Party System,* 65, 73, 103, 108; Abbott, "Massachusetts," 3–5, 9; and Coyne, "Hoosac Tunnel," 229.

24. *Boston Post,* June 6, 8, 9, 1865.

25. Schexnayder, *Builders of the Hoosac Tunnel,* 306, 328.

26. *Scribner's Magazine* (Dec. 1870), 151; and Coyne, "Hoosac Tunnel," 240–246.

27. One of the assistant engineers at the tunnel, W. P. Granger, was transporting tri-nitro-glycerin over the Hoosac Mountain when his wagon turned over and the explosive froze solid in a snow bank. He subsequently discovered that it could not be ignited when in a frozen state. See *Report on the Hoosac Tunnel . . . by the Joint Standing Committee of 1866,* 39–40; *Adams Transcript,* Dec. 29, 1870; Dalrymple, *History of the Hoosac Tunnel,* 9–10; Byron, *Pinprick of Light,* 44–46; and Coyne, "Hoosac Tunnel," 242–245, 251.

28. *Report on the Hoosac Tunnel . . . by the Joint Standing Committee of 1866,* 25–27; Brierley, "Construction of the Hoosac Tunnel," 201–202; and Byron, *Pinprick of Light,* 42–44.

29. *Report on the Hoosac Tunnel . . . by the Joint Standing Committee of 1866,* 23–24; and Schexnayder, *Builders of the Hoosac Tunnel,* 306–308.

30. Byron, *Pinprick of Light,* 49, 50; and Schexnayder, *Builders of the Hoosac Tunnel,* 315–317.

6. DISASTER AND RECKONING

1. The opening paragraph of this chapter was constructed from newspaper accounts, photographs from the period, and the author's visit to the site of the central shaft tragedy on the 150th anniversary of the event. The Hoosac Tunnel scrapbooks at the North Adams Public Library were especially useful. See also *Boston Post,* Oct. 21, 23, 1867, and *Boston Herald,* Oct. 23, 1867.

2. *Adams Transcript,* April 11, Aug. 15, Oct. 24, 1867; Byron, *Pinprick of Light,* 49–51; Schexnayder, *Builders of the Hoosac Tunnel,* 318–319; and Dalrymple, *History of the Hoosac Tunnel,* 14.

3. *Boston Post,* Oct. 23, 1867; Byron, *Pinprick of Light,* 51–52; and Coyne, "Hoosac Tunnel," 261–262.

4. *Boston Herald,* Oct. 21, 1867; *Boston Post,* Oct. 23, 1867; *Adams Transcript,* Oct. 24, 1867; and Coyne, "Hoosac Tunnel," 260–261.

5. *Boston Post,* Oct. 21, 23, 1867; *Adams Transcript,* Oct. 24, 1867; and *Scientific American* (Nov. 9, 1867), Vol. XVII, No. : 19, 293.

6. Governor Bullock annulled the contract with Dull, Gowan & White on Nov. 9, 1867. It is unclear who paid for the equipment and materials lost in the central shaft fire. See Coyne, "Hoosac Tunnel," 264–265; Schexnayder, *Builders of the Hoosac Tunnel,* 322, 329; *Report of Benjamin H. Latrobe, Consulting Engineer, on the Troy and Greenfield Railroad and Hoosac Tunnel* (Boston: Wright & Potter, 1869), 16; and Browne, *Hoosac Tunnel Days,* 6–7.

7. *Report of Benjamin H. Latrobe,* 18; and Kirkland, "Hoosac Tunnel," 107–108.

8. It is in *Richard III,* Act 4, Scene 4, where Queen Elizabeth calls out, "Where is Clarence? Where is gentle Rivers, Vaughan, Gray? Where is kind Hastings?" See F. W. Bird, *The Modern Mi-*

notaur (Boston: J. E. Farwell, 1868), 3, 23, 37–38; and F. W. Bird, *The Last Agony of the Great Bore* (Boston: E. P. Dutton, 1868), 7, 22, 62–63.

9. Coyne, "Hoosac Tunnel," 270–275; Bird, *Last Agony*, 63–65; and Abbott, "Massachusetts," 11.

10. The author of the *The Death of Our Minotaur* was Edward Hamilton, whose name is written on a copy of the pamphlet at the Massachusetts Historical Society. Clearly an accomplished satirist, Hamilton was probably paid handsomely by the tunnelites for his work. The word "Pecksniffian," meaning hypocritical or falsely moral, is derived from the unctuous hypocrite, Pecksniff, in Charles Dickens's novel *Martin Chuzzlewit* (1843). See Edward Hamilton, *The Death of Our Minotaur* (Boston: publisher unknown, 1868), 3–4, 9–10, 13; Baum, *Civil War Party System*, 5, 152, 211, 215; and Coyne, "Hoosac Tunnel," 269–270.

11. Coyne, "Hoosac Tunnel," 277–278, 280.

12. Baum, *Civil War Party System*, 123–125, 130–131; Abbott, "Massachusetts," 9–10; Coyne, "Hoosac Tunnel," 280–283; and *Boston Post*, Nov. 31, 1868.

13. It is unclear how much the public understood about progress at the Hoosac Tunnel. The Commissioners' Report for the year 1868 explained that only 9,338 feet of the tunnel's intended 25,031-foot length had been excavated (37 percent). Other details included 931 feet of brick arching at the west end and 583 feet of the 1,028 foot central shaft (56 percent). See *Report of the Commissioners Upon the Troy and Greenfield Railroad and Hoosac Tunnel* (Boston: Wright & Potter, 1869), 27; *Boston Post*, Jan. 8, 1869; F. W. Bird, *The Hoosac Tunnel Contract* (Boston: Wright & Potter, 1869), 9, 23, 18; Coyne, "Hoosac Tunnel," 283–285; Schexnayder, *Builders of the Hoosac Tunnel*, 344; Abbott, "Massachusetts," 10–11; and Austin, *History of Massachusetts*, 535.

14. *Boston Post*, Jan. 8, 1869.

15. Governor Claflin's veto of state aid to the Boston, Hartford & Erie Railroad effectively ended the life of this line. The railroad was badly managed and its stock price depressed. Still, Claflin's veto cost him votes in parts of Massachusetts supporting the railroad. In 1868, President Andrew Johnson was impeached in the U.S. House of Representatives. However, the trial in the U.S. Senate to remove him from office failed by one vote. See Austin, *History of Massachusetts*, 536; *Boston Post*, Jan. 6, 1869; Baum, *Civil War Party System*, 138–144; Abbott, "Massachusetts," 12–13; and Richard White, *The Republic for Which It Stands: The United States during Reconstruction and the Gilded Age, 1865–1896* (New York: Oxford UP, 2017), 91–96.

16. *Boston Post*, Jan. 11, 1869; and Perry Lewis, *Boats against the Current: American Culture between Revolution and Modernity, 1820–1860* (Lanham, MD: Rowman & Little, 1993), 8, 135–136.

17. Cliff Schexnayder has extensively researched the background of the Shanly brothers. See Schexnayder, *Builders of the Hoosac Tunnel*, 349–385.

18. *Adams Transcript*, Jan. 21, 1869, and June 23, 1870.

19. *Joint Standing Committee of 1868, on the Hoosac Tunnel and Troy and Greenfield Railroad (Feb. 16, 1869)* (Boston: Wright & Potter, 1869), 4, 6, 12; Schexnayder, *Builders of the Hoosac Tunnel*, 387–388, 390; and Coyne, "Hoosac Tunnel," 288.

20. Coyne, "Hoosac Tunnel," 291–292; Kirkland, "Hoosac Tunnel," 108; and Schexnayder, *Builders of the Hoosac Tunnel*, 395.

21. Coyne, "Hoosac Tunnel," 294–296; Byron, *Pinprick of Light*, 61- 63; "Report of James Lau-

rie, Consulting Engineer" in *Senate—No. 283* (Boston: publisher unknown, 1871), 20–21; and *Adams Transcript,* Oct. 7, 1869.

22. The estimated damage from the flood was $18,000. The Shanlys would eventually be reimbursed by the state but only after six years of petitions and hearings. The railroad from Greenfield to the east portal remained out of commission for nine months, because the state had no funds authorized for repairs. See Coyne, "Hoosac Tunnel," 295–296.

23. Efforts to merge the Boston & Worcester and the Western Railroad had failed in 1846 and 1855. It was the recognition that the Hoosac Tunnel would soon be completed that finally spurred stockholders in the two railroads to consolidate. See *Boston Post,* Feb. 23, 1866; Salsbury, *State, Investor, and Railroad,* 291, 293–294, 302; Kirkland, *Men, Cities, and Transportation,* Vol. 1, 366–368; Austin, *History of Massachusetts,* 528; and Hill, *Boston's Trade and Commerce,* 12–13.

24. The immediate reason for the break between President Grant and Senator Charles Sumner was the proposed annexation of Santo Domingo (the modern Dominican Republic). Grant became fixated on the annexation of the island, and Sumner opposed it. When Grant succeeded in having Sumner removed from the chairmanship of the Senate Foreign Relations Committee, Radical Republicans turned on Grant. See White, *Republic for Which It Stands,* 202–205; and Baum, *Civil War Party System,* 147–150, 173–176, 183.

25. Massachusetts was by no means in the vanguard with its railroad commission. Several New England states already had them. Furthermore, these commissions were largely advisory. Commissions with real power would first emerge in the West, where farmers had issues with unfair shipping rates and grain elevator prices. Broad railroad regulation would have to wait for the Interstate Commerce Act in 1887. See Stover, *American Railroads,* 117–122; Abbott, "Massachusetts," 11–12; and Charles Francis Adams, Jr., and Henry Adams, *Chapters of Erie and Other Essays* (New York: Henry Holt, 1891), 1.

26. The arguments over surveying and grading show up in legislative records. For example, see *Senate—No. 330* (Boston: publisher unknown, 1871), 2–9; and Coyne, "Hoosac Tunnel," 300–305.

27. Shexnayder, *Builders of the Hoosac Tunnel,* 411, 413.

28. Coyne, "Hoosac Tunnel," 305–306; and Dalrymple, *History of the Hoosac Tunnel,* 12–13.

29. Byron, *Pinprick of Light,* 67–68; Coyne, "Hoosac Tunnel," 305–309; and Schexnayder, *Builders of the Hoosac Tunnel,* 414–416.

30. It is likely that Otis designed the whole elevator, since he had installed a smaller one in the tunnel's west shaft a few years earlier. See *Adams Transcript,* Oct. 20, 1870; and Byron, *Pinprick of Light,* 66–67.

31 Coyne, "Hoosac Tunnel," 312; Schexnayder, *Builders of the Hoosac Tunnel,* 430–432; and Browne, *Hoosac Tunnel Days,* 14.

32. Coyne, "Hoosac Tunnel," 316–322.

33. Coyne, "Hoosac Tunnel," 320–322.

34. Austin, *History of Massachusetts,* 483; and White, *Republic for Which It Stands,* 260–262.

35. Charles Cahoon has documented 195 accidents at the tunnel ending in death or serious injury. See Cahoon, "Hoosac Tunnel Accident Victims."

36. "Only a Tunneller" was written by E. A. Wright. See Dalrymple, *History of Hoosac Tunnel,* 17; *Adams Transcript,* March 12, 1874; and *Hoosac Valley News,* Nov. 26, 1873.

7. BREAKTHROUGH AND PANIC

1. As miners tunneling east from the west shaft approached those coming from the west heading of the central shaft, the *Adams Transcript* published progress reports at the start of every month. See Coyne, "Hoosac Tunnel," 325–326; and Schexnayder, *Builders of the Hoosac Tunnel,* 449, 457–459.

2. White, *Republic for Which It Stands,* 261–263, 266–268; Rezneck, *Business Depressions,* 129–131; and Stover, *American Railroads,* 97, 114.

3. Kirkland, "Hoosac Tunnel," 108; Schexnayder, *Builders of the Hoosac Tunnel,* 436–437, 441; and Byron, *Pinprick of Light,* 64–65.

4. The first people to go through the five-foot hole connecting the central shaft and east portal were two local school boys named Frank Blanchard and Dallas Dean. They had disguised themselves as miners and ridden down the central shaft unnoticed. See *Senate—No. 283* (Boston: publisher unknown, 1873), 25–26; Byron, *Pinprick of Light,* 71; Schexnayder, *Builders of the Hoosac Tunnel,* 441, 436–437; Coyne, "Hoosac Tunnel," 318, 322; and Wheelwright, *Alvah Crocker,* 52.

5. Engineer Benjamin Frost took full credit for the perfect alignment of the headings in the first breakthrough. See *Senate—No. 201* (Boston: publisher unknown, 1873), 6–7; and Coyne, "Hoosac Tunnel," 325–326.

6. The Lowell & Boston was the first modern railroad in Massachusetts, chartered in 1830 and finished in 1835. In its early life, it served Lowell textile mills and, later, connected the Bay State to Portland, Maine. See Stover, *American Railroads,* 14–15; Kirkland, "Hoosac Tunnel," 110–112; Coyne, "Hoosac Tunnel," 334–335; *Report of the Second Hearing on the Hoosac Tunnel before the Committee on Railroads, Jan. 30, 1873* (Boston: Wright & Potter, 1873), 16–19; *Report on Fourth Hearing . . . Feb. 9, 1873* (Boston: Wright & Potter, 1873), 24–27; *Report on the Seventh Hearing . . . Feb. 13, 1873* (Boston: Wright & Potter, 1873), 23, 25, 28–29; and *Report on Eighth Hearing . . . Feb. 14, 1873* (Boston: Wright & Potter, 1873), 4–10.

7. One exacerbating factor in the hearings was the rumor that New York railroad tycoon Cornelius Vanderbilt had designs on the Hoosac Tunnel. See *Minority Report of the Committee on Railroads in Relation to the Hoosac Tunnel by Hon. E. P. Carpenter* (Boston: Wright & Potter, 1873), 5–7, 31–35; *Consolidation of the Tunnel Line: Closing Arguments of Benj. F. Thomas* (Boston: Wright & Potter, 1873), 5–7; Edward Atkinson, *How to Pay for the Hoosac Tunnel* (Boston: publisher unknown, 1873), 5, 11; Coyne, "Hoosac Tunnel," 328–329, 342–351; *Fourteenth Hearing . . . Mar. 13, 1873* (Boston: Wright & Potter, 1873), 18, 22–23, 28; *Fifteenth Hearing . . . Mar. 14, 1873* (Boston: Wright & Potter, 1873), 11–12; *Sixteenth Hearing . . . Mar. 15, 1873* (Boston: Wright & Potter, 1873), 7–8, 14; and *Nineteenth Hearing . . . Mar. 18, 1873* (Boston: Wright & Potter, 1873), 24–27.

8. John Stover has described the antebellum years as a period of "degeneration" for the American railroad industry. See Stover, *American Railroads,* 96–87, 102–111; Dobbin and Dowd, "How Policy Shapes Competition," 506; White, *Republic for Which It Stands,* 255–260; Lears, *Rebirth of a Nation,* 54–55; T. J. Stiles, *The First Tycoon: The Epic Life of Cornelius Vanderbilt* (New York: Vintage, 2017), 467–470; Peter Knight, *Reading the Market: Genres of Financial Capitalism in Gilded Age America* (Baltimore: Johns Hopkins UP, 2016), 197–198; and Adams, *Chapters of Erie,* 10, 22, 43.

9. Jay Cooke & Company failed on September 13, 1873. A week later, a full-fledged panic had

set in. The Panic of 1873 had its origins in Europe. German banks, flush with reparations payments after the Franco-Prussian War, invested heavily in Austrian banks. The Austrian banks, in turn, lent too liberally to grain farmers in the Austro-Hungarian Empire but suffered when cheaper American wheat began to dominate the market. When the Austrian banks failed, the Bank of England raised interest rates. Highly dependent on British credit, the American economy began to contract from a lack of liquidity. In this sense, the causation of the Panic of 1873 was circular. Because they were so leveraged, railroads were especially vulnerable to a drying up of credit. See Rezneck, *Business Depressions*, 129, 130, 131; and White, *Republic for Which It Stands*, 260–262.

10. Rezneck, *Business Depressions*, 131–132.

11. Wheelwright, *Alvah Crocker*, 53; and Byron, *Pinprick of Light*, 72–74.

12. *Adams Transcript*, Dec. 26, 1872; Byron, *Pinprick of Light*, 72–76; and Schexnayder, *Builders of the Hoosac Tunnel*, 458–460.

13. *Adams Transcript*, Dec. 4, 1873.

14. Schexnayder, *Builders of the Hoosac Tunnel*, 462–463.

15. Byron, *Pinprick of Light*, 72.

16. Schexnayder, *Builders of the Hoosac Tunnel*, 475–476, 491; and *Report of the Commissioners . . . February 28, 1863* (Boston: Wright & Potter, 1863), 39–40.

17. Byron, *Pinprick of Light*, 83–85.

18. Wheelwright, *Alvah Crocker*, 73.

19. *Senate Bill—No. 111* (Boston, 1873), 12, 19–20; Schexnayder, *Builders of the Hoosac Tunnel*, 470–477; Byron, *Pinprick of Light*, 76; Geoffrey Blodgett, *The Gentle Reformers: Massachusetts Democrats in the Cleveland Era* (Cambridge, MA: Harvard UP, 1966), 555–558; and "Notes of A. E. Bond" (unpaginated notes in the holdings of the North Adams Public Library).

20. Rezneck, *Business Depressions*, 130, 134–139, 146; and White, *Republic for Which It Stands*, 267–271.

21. Francis Bird ran for governor of Massachusetts on the Liberal Republican–Democratic alliance ticket in 1872. The *Adams Transcript* reminded readers of his opposition to the Hoosac Tunnel and urged them to vote for his opponent, William Washburn, the Republican governor up for reelection. Bird lost badly to Washburn. See *Adams Transcript*, Oct. 24, 1872; White, *Republic for Which It Stands*, 274–275; Baum, *Civil War Party System*, 183–194; and Abbott, "Massachusetts: Maintaining Hegemony," 9–10, 21.

22. The Chimera is a hybrid animal from Greek mythology that has come to symbolize illusion, fabrication, or an unrealized dream, all of which seem to fit the Hoosac Tunnel. The mist-and-fog metaphors are borrowed from Edward Kirkland. See Kirkland, *Men, Cities, and Transportation*, Vol. 1, 398.

23. *Boston Post*, June 5, 8, 1865, Jan. 11, 1868; Ralph Waldo Emerson, *The Complete Works of Ralph Waldo Emerson* (Boston: Houghton Mifflin, 1904), Vol. 1, 365; Sellers, *Market Revolution*, 380–381; and Smith, *Virgin Land*, 95, 185.

24. Edward Kirkland argued that Boston financiers could easily have bought the New York Central during the 1850s or early 1860s. Kirkland also explained that Vanderbilt earned $40.00 on each ten-ton rail car of wheat from Chicago to New York but only $25.83 if it was transferred at Albany to Boston. See Kirkland, *Men, Cities, and Transportation*, Vol. 1, 154–155, 373; Stover,

American Railroads, 40, 101–102; Wolmar, *Railroad Revolution*, 108, 240, 241; Johnson and Supple, *Boston Capitalists and Western Railroads*, 177–180; Salsbury, *State, Investor, and Railroad*, 302–303; Stiles, *First Tycoon*, 403, 407, 453, 482, 495–496, 506; and William Cronon, *Nature's Metropolis: Chicago and the Great West* (New York: W. W. Norton, 1991), 109–113, 145.

25. It was noteworthy that Elias Hasket Derby highlighted the deficiencies of Boston's harborside facilities during the Great Debate of 1873. Derby was an acknowledged expert on the city's economy and infrastructure. See Coyne, "Hoosac Tunnel," 335; Kirkland, *Men, Cities, and Transportation*, Vol. 1, 523–524; Stiles, *First Tycoon*, 500–501; and Edwin G. Burrows and Mike Wallace, *Gotham: A History of New York City to 1898* (New York: Oxford UP, 1999), 436, 654–656.

26. Given that a large component of wheat pricing was driven by transportation cost, the saturation of railroad lines in America's wheat-growing area and low cost of wheat shipments from Chicago to New York brought the country's wheat to competitive lows. What is more, the cost of New York to Liverpool freight fell by two-thirds between 1825 and 1855. By 1870, American wheat beat the cost of Eastern European wheat landed at Liverpool. Edward Collins established a New York–Liverpool steamship link, the New York Mail Steamship Company, in 1848. Although Collins received a generous mail-carrier subsidy from the U.S. government, he never made money on the line. New York's Ocean Steam Navigation Company was also going to Bremen during this period. See Burrows and Wallace, *Gotham*, 433, 647–650, 654; Taylor, *Transportation Revolution*, 115–121, 147–148; Hill, *Boston Trade and Commerce*, 13–14; North, *Economic Growth*, 150–151, 207, 210–211; Stiles, *First Tycoon*, 104–110, 256, 265–268; Morison, *Maritime History*, 235, 255, 339, 336; and Kirkland, *Men, Cities, and Transportation*, Vol. 1, 20–25, 174, 183.

27. McKey, "Elias Hasket Derby," 155, 158; Dalzell, *Enterprising Elite*, 31–32; and Dolin, *Leviathan*, 109–110.

28. The rise of New York City to commercial ascendency is too complex to fully describe here. In addition to the factors already discussed, the establishment of investment banking houses, trust companies, wholesaler merchant firms, and the New York Stock and Exchange Board were extremely important. Finance houses like Brown Brothers that extended credit to southern planters and bought their cotton played a major role. See Burrows and Wallace, *Gotham*, 436, 445, 570; Beckert, *Empire of Cotton*, 217–220; Johnson and Supple, *Boston Capitalists and Western Railroads*, 177–180, 188–191; Stiles, *First Tycoon*, 19, 40–41, 78, 263, 330; and Russell B. Adams, Jr., *The Boston Money Tree* (New York: Thomas Crowell, 1977), 102, 137–140, 189.

29. *Adams Transcript*, Feb. 11, April 8, 1875; Byron, *Pinprick of Light*, 80–82; and Schexnayder, *Builders of the Hoosac Tunnel*, 482–484.

30. The railroad described here was the Boston, Hoosac Tunnel & Western. Its Boston investors were Oliver and F. L. Ames, Elisha Atkins, and William Burt. In addition to problems with the Massachusetts legislature, the New York court system opposed the venture and blocked its progress. So, too, the Panic of 1873 created havoc with its financing. This author wonders whether the five "corporators," especially Charles Francis Adams, Jr., whose railroad expertise was widely respected, might not have demonstrated more leadership in the 1875 hearings. Their report to the legislature was weak in its overall recommendations and specific actions to be taken. See *Report of the Corporators, January 1875* (Boston: Wright & Potter, 1875), 12–14, 21, 24, 27; Kirkland, *Men, Cities, and Transportation*, Vol. 1, 418–422; and Austin, *History of Massachusetts*, 588.

31. *Mr. Bowman's Bill: Some Reasons for its Adoption* (Boston: Wright & Potter, 1875), 2–3; and Kirkland, *Men, Cities, and Transportation*, Vol. 1, 112.

32. *Boston Evening Transcript*, Feb. 20, 21, 28, March 1, 1877; *Report of the Joint Committee on the Hoosac Tunnel and the Troy and Greenfield Railroad April 1877* (Boston: Albert J. Wright, 1877), 4, 10, 18; and Kirkland, *Men, Cities, and Transportation*, Vol. 1, 422–423.

33. With the completion of the east and west facades, the Hoosac Tunnel reached its full length of 25,081 feet, or 4.75 miles. Under the toll-gate plan, the state owned the tunnel and the forty-four miles of the Troy & Greenfield Railroad. See Kirkland, *Men, Cities, and Transportation*, Vol. 1, 423–425; *Report of the Joint Committee . . . April 1877*, 6–7, 18; and Byron, *Pinprick of Light*, 79–82.

34. White, *Republic for which It Stands*, 271–272, 479; Lears, *Rebirth of a Nation*, 75–76; Rezneck, *Business Depressions*, 132–136; and Austin, *History of Massachusetts*, 564, 568–569, 571.

35. Levine, *Power of Darkness*, 192–193; Andrew Delbanco, *Melville: His World and Work* (New York: Vintage, 2005), 246–250; Knight, *Reading the Market*, 146–149; and Herman Melville, *The Confidence-Man* (1857; rpr. New York: Signet, 1954), 12, 15, 21, 55, 73, 98, 122–123, 162, 119–120, 122–123, 205.

36. Louis Menand has argued that the United States became a different place after the Civil War. Drew Gilpin Faust has observed that literature was incapable of describing the war and its horrors. The disputed presidential election of 1876 was between Republican Rutherford Hayes and Democrat Samuel Tilden. The latter won the popular vote but a political compromise gave deciding electoral votes in the South to Hayes. After Hayes took office, military troops were withdrawn from South Carolina, Louisiana, and Florida. Given the complexity of the compromise, historians debate the precise impact of the compromise on the end of Reconstruction. The Beecher-Tilton trial captured public attention during 1875 as few so-called sex scandals have before or since. Henry Ward Beecher was minister of the Plymouth Church in Brooklyn, New York, and a highly respected national religious figure. See Menand, *Metaphysical Club*, ix, x, xii; Drew Gilpin Faust, *This Republic of Suffering: Death and the American Civil War* (New York: Knopf, 2008), 208–209; Lears, *Rebirth of a Nation*, 23, 54–56; Van Wyck Brooks, *The Times of Melville and Whitman* (New York: Dutton, 1947), 234–235; Michael F. Holt, *By One Vote: The Disputed Election of 1876* (Lawrence: Univ. of Kansas Press, 2008), 239–242, 246–247; Debby Applegate, *The Most Famous Man in America: The Biography of Henry Ward Beecher* (New York: Doubleday, 2006), 442–449; and White, *Republic for Which It Stands*, 162–164, 261, 269, 320–337, 444–445, 525, 571, 855.

37. *Adams Transcript*, Oct. 2, Dec. 4, 1873.

38. These exports include wheat, flour, corn, and corn meal and are from Hamilton Andrews Hill. See "Report from J. Prescott, Manager" in *Senate—No. 44* (Boston: publisher unknown, 1877), 2, 6, 11–14; and Hill, *Boston's Trade and Commerce*, 18.

39. The State of Massachusetts had already sold off the last of its stock in the Boston & Albany Railroad. Reform-minded Republicans now turned to the state's much larger commitment to the Troy & Greenfield Railroad. See Kirkland, *Men, Cities, and Transportation*, Vol. 1, 425–426; and *Twelfth Annual Report of the Manager of the Troy & Greenfield Railroad and the Hoosac Tunnel for the Year ending Sept. 30, 1886* (Boston: publisher unknown, 1887), 6, 9, 16.

40. Interestingly, Herman Haupt still owned 12,372 shares of worthless Troy & Greenfield

stock. However, the state could not sell the railroad to the Fitchburg until Haupt sold his stock back to the state. Haupt ultimately settled for $105,608, or eight cents on the dollar. See Kirkland, *Men, Cities, and Transportation,* Vol. 1, 425–426; and Byron, *Pinprick of Light,* 83.

41. General William L. Burt was a Civil War veteran and an ex-postmaster of Boston. He led the consortium of local investors behind the Boston, Hoosac Tunnel & Western Railroad. He not only vilified the Troy & Boston Railroad as a decrepit road over which to enter New York, but also called the Fitchburg Railroad a mismanaged line, incapable of fulfilling the Hoosac Tunnel dream. Ironically, the Fitchburg ended up buying Burt's railroad. Given the price received, he was probably not disappointed in this outcome. See Kirkland, *Men, Cities, and Transportation,* Vol. 1, 421–422, 424, 427–428; and *Closing Arguments of Gen. William L. Burt, before the Joint Special Committee of the Legislature on the Hoosac Tunnel and Troy & Greenfield Railroad, March 1877* (Boston: Albert J. Wright, 1877), 160, 162, 169, 195.

42. *Fitchburg Railroad Company. The Port of Boston, Hoosac Tunnel Docks, and Elevators* (Boston: publisher unknown, 1882), no pagination.

43. Railroad regulation at the state level came out of the Granger movement in the West, where farmers in states like Iowa and Wisconsin objected to high freight rates for shipping their produce and high grain elevator fees to store it. These so-called Granger Laws were followed by passage of the Interstate Commerce Act in 1887. For the first fifteen years of its life, this act remained relatively benign. The only late railroad expansion in New England occurred in Maine, where the Aroostook Railroad accessed northern spruce for paper production. See Stover, *American Railroads,* 122–125, 136.

44. The Boston & Maine Railroad was chartered in New Hampshire in 1835. It leased the Fitchburg Railroad in 1900 and bought it outright in 1919. During the 1960s and 1970s, the Boston & Maine declined with the rest of the railroad industry. See Kirkland, *Men, Cities, and Transportation,* Vol. 1, 429–430, 463; and Stover, *American Railroads,* 195–197, 220–225.

CONCLUSION

1. *Report of the Commissioners,* 39–40, 49–52.

2. Numerous estimates have been made of the Hoosac Tunnel's cost. The most credible is that of Edward Chase Kirkland, cited here. See Kirkland, *Men, Cities, and Transportation,* 431- 432; *Fourteenth Hearing . . . March 13, 1873,* 22; *Hoosac Tunnel: A Brief Report,* 67; and Atkinson, *How to Pay for the Hoosac Tunnel,* 5, 11.

3. The best statistics for Boston's western exports are from 1881. Boston exported 18,777, 290 bushels of wheat, flour, and corn that year. The value of these commodities was $72,100,198. See Hill, *Boston's Trade and Commerce,* 18; and Kirkland, *Men, Cities, and Transportation,* 430–431.

4. Boston never seriously rivaled New York in the export of wheat, flour, and corn. In 1881, New York exported an extraordinary 121,284,878 bushels of these commodities. Boston exported only 15 percent of this volume. Boston did, however, grow its share of exports at the expense of Philadelphia, Baltimore, and New Orleans during the 1880s and 1890s. The Shanly brothers had worked on the Grand Trunk railroad in the 1850s. Walter Shanly became general manager and

chief engineer of the railroad in 1858. See Kirkland, *Men, Cities, and Transportation*, 516–517, 521, 524–525; Stover, *American Railroads*, 122–124; and Schexnayder, *Builders of the Hoosac Tunnel*, 378.

5. Kirkland, *Men, Cities, and Transportation*, 522–525.

6. Hill, *Boston's Trade and Commerce*, 18; and North, *Economic Growth*, 42–43.

7. The five-mile Connaught Tunnel in the Canadian Rockies was completed in 1916, followed by the 6.2-mile Moffat Tunnel west of Denver in 1927. In 1929, the 7.8-mile Cascade Tunnel in Washington State became the longest tunnel in North America. It was exceeded by the 9.1-mile MacDonald Tunnel in the Canadian Rockies in 1988. See Byron, *Pinprick of Light*, 76–77; and *North Adams Transcript*, May 18, 1927.

SELECTED BIBLIOGRAPHY

The following bibliography does not include the Massachusetts State House documents fully described in this book's endnotes. All were published in Boston by the state printer at the time and are held at the Massachusetts Historical Society, American Antiquarian Society, North Adams Public Library, or the Massachusetts State House Archives.

Abbott, Richard. "Massachusetts: Maintenance and Hegemony," in James C. Mohr, ed., *Radical Republicans in the North: State Politics during Reconstruction*. Baltimore: Johns Hopkins UP, 1976.

Achenbach, Joel. *The Grand Idea: George Washington's Potomac and the Race to the West*. New York: Simon & Schuster, 2004.

Adams, Charles Francis, Jr., and Henry Adams. *Chapters of Erie and Other Essays*. New York: Henry Holt, 1891.

Adams, Russell B., Jr. *The Boston Money Tree*. New York: Thomas Crowell, 1977.

Anbinder, Tyler. *Nativism & Slavery: The Northern Know Nothings & The Politics of the 1850s*. New York: Oxford UP, 1997.

Applegate, Debby. *The Most Famous Man in America: The Biography of Henry Ward Beecher*. New York: Doubleday, 2006.

Austin, George Lowell. *The History of Massachusetts*. Boston: B. B. Russell, 1884.

Bailyn, Bernard. *The Barbarous Years: The Peopling of British North America and the Conflict of Civilizations, 1600–1675*. New York: Knopf, 2012.

Baltzell, E. Digby. *Puritan Boston and Quaker Philadelphia*. New York: Free Press, 1979.

Barnes, Gilbert Hobbs. *The Anti-Slavery Impulse, 1830–1844*. New York: Harcourt Brace, 1933.

Basbanes, Nicholas A. *On Paper: The Everything of Its Two-Thousand-Year History*. New York: Vintage, 2014.

Baum, Dale. *The Civil War Party System: The Case of Massachusetts, 1848–1876*. Chapel Hill: University of North Carolina Press, 1984.

Beckert, Sven. *Empire of Cotton: A Global History*. New York: Vintage, 2014.

Bensel, Richard Franklin. *Yankee Leviathan: The Origins of Central State Authority in America, 1859–1877*. New York: Cambridge UP, 1990.

Berns, John F. *Providence & Patriotism in Early America, 1640–1815*. Charlottesville: University of Virginia Press, 1978.

Bernstein, Peter L. *Wedding of the Waters: The Erie Canal and the Making of a Great Nation*. New York: W. W. Norton, 2005.

Bird, F. W. *The Road to Ruin, or, The Decline and Fall of the Hoosac Tunnel*. Boston: Wright & Potter, 1862.

———. *The Hoosac Tunnel: Our Financial Maelstrom*. Boston: E. P. Dutton, 1866.

———. *The Last Agony of the Great Bore*. Boston: E. P. Dutton, 1868.

———. *The Modern Minotaur*. Boston: J. E. Farwell, 1868.

———. *The Hoosac Tunnel Contract*. Boston: Wright & Potter, 1869.

Black, Andrew R. *John Pendleton Kennedy: Early American Novelist, Whig Statesman & Ardent Nationalist*. Baton Rouge: Louisiana State UP, 2016.

Blodgett, Geoffrey. *The Gentle Reformers: Massachusetts Democrats in the Cleveland Era*. Cambridge, MA: Harvard UP, 1966.

Booth, Robert. *Death of an Empire: The Rise and Murderous Fall of Salem, America's Richest City*. New York: St. Martin's, 2011.

Brierley, Gary S. "Construction of the Hoosac Tunnel, 1855 to 1876," in *Journal of Boston Civil Engineers Section, American Society of Civil Engineers*, 63.3 (Oct. 1976): 184–202.

Brooks, Van Wyck. *The Times of Melville and Whitman*. New York: Dutton, 1947.

Browne, Isaac S. *Hoosac Tunnel Days: Accidents and Accident Victims, 1859–1878*. North Adams, MA: publisher unknown, undated.

Browne, William B. "Chamber of Commerce Speech, March 4, 1924." North Adams, MA: publisher unknown, 1924.

Burrows, Edwin G., and Mike Wallace. *Gotham: A History of New York City to 1898*. New York: Oxford UP, 1999.

Burstein, Andrew. *America's Jubilee: A Generation Remembers the Revolution after 50 Years of Independence*. New York: Vintage, 2007.

Byron, Carl R. "Hoosac Tunnel: The Mohawk Trail by Rail," in *B & M Bulletin* (Boston & Maine) 3.1 (Sept. 1973): 1–14.

———. *A Pinprick of Light: The Troy and Greenfield Railroad and Its Hoosac Tunnel*. Shelburne, VT: New England Press, 1974.

Conners, Anthony J. *Ingenious Machinists: Two Inventive Lives from the American Industrial Revolution*. Albany: State University of New York Press, 2014.

Coyne, Terrence Edward. "The Hoosac Tunnel." PhD dissertation, Clarke University, 1992.

———. "The Hoosac Tunnel: Massachusetts' Western Gateway," in *Historical Journal of Massachusetts* 23.1 (Winter 1995): 1–24.

Cressy, David. *Coming Over: Migration and Communication between England and New England in the Seventeenth Century.* New York: Cambridge UP, 1987.

Cronon, William. *Changes in the Land: Indians, Colonists, and the Ecology of New England.* New York: Hill & Wang, 1987.

———. *Nature's Metropolis: Chicago and the Great West.* New York: W. W. Norton, 1991.

Dalrymple, Oscar. *History of the Hoosac Tunnel.* North Adams, MA: North Adams Historical Society, 1880.

Dalzell, Robert F., Jr. *Enterprising Elites: The Boston Associates and the World They Made.* Cambridge, MA: Harvard UP, 1987.

Dangerfield, George. *The Awakening of American Nationalism, 1815–1828.* New York: Harper & Row, 1965.

Daniels, Roger. *Coming to America: A History of Immigration in American Life.* New York: Harper Collins, 1990.

Delbanco, Andrew. *The Puritan Ordeal.* New York: Charles Scribner, 1989.

———. *Melville: His World and His Work.* New York: Vintage, 2005.

Derby, Elias Hasket. *Two Months Abroad: Or, a Trip to England, France, Baden, Prussia, and Belgium in August and September, 1843.* 1843. Reprint, Columbia, SC: Wentworth, 2016.

Dilts, James D. *The Great Road: The Building of the Baltimore & Ohio, the Nation's First Railroad, 1828–1853.* Stanford, CA: Stanford UP, 1993.

Dobbin, Frank, and Timothy J. Dowd. "How Policy Shapes Competition: Early Railroad Foundings in Massachusetts," in *Administration Science Quarterly* 42.3 (1997): 501–529.

Dolin, Eric Jay. *Leviathan: The History of Whaling in America.* New York: W. W. Norton, 2007.

———. *When America First Met China: An Exotic History of Tea, Drugs, and Money in the Age of Sail.* New York: W. W. Norton, 2012.

Drabble, Margaret, ed. *Oxford Companion to Literature.* New York: Oxford UP, 1985.

Elliot, J. H. *Empires of the Atlantic World: Britain and Spain in America, 1492–1830.* New Haven, CT: Yale UP, 2006.

Emerson, Ralph Waldo. *The Complete Works of Ralph Waldo Emerson,* Vol. 1. Boston: Houghton Mifflin, 1904.

———. *The Complete Works of Ralph Waldo Emerson: Natural History of Intellect and Other Papers,* Vol. 12. Ann Arbor: University of Michigan Press, 2006.

Faust, Drew Gilpin. *This Republic of Suffering: Death and the American Civil War.* New York: Knopf, 2008.

Foner, Eric. *The Fiery Trial: Abraham Lincoln and American Slavery.* New York: W. W. Norton, 2010.

Formisano, Ronald P. *The Transformation of Political Culture: Massachusetts Parties, 1790–1840s.* New York: Oxford UP, 1983.

Francis William Bird: A Biographical Sketch. Boston: Norwood, 1897.

Fredrickson, George M. *The Inner Civil War: Northern Intellectuals and the Crisis of the Union.* New York: Harper & Row, 1965.

Garraty, John A., and Mark C. Carnes, eds. *American National Biography*, Vol. 2. New York: Oxford UP, 1999.

Gienapp, William E. *The Origins of the Republican Party, 1852–1856.* New York: Oxford UP, 1987.

Hall, David D. *A Reforming People: Puritans and the Transformation of Public Life in New England.* New York: Knopf, 2001.

Hamilton, Edward. *Death of Our Minotaur.* Boston: publisher unknown, 1868.

Handlin, Oscar. *Boston Immigration.* Cambridge, MA: Belknap Press, 1979.

Harrington, Fred H. *Fighting Politician: Major General N. P. Banks.* Philadelphia: University of Pennsylvania Press, 1948.

Hartford, William F. *Money, Morals, and Politics: Massachusetts in the Age of the Boston Associates.* Boston: Northeastern UP, 2001.

Haupt, Herman. *Reminiscences of General Herman Haupt.* Milwaukee: Wright & Joy, 1901.

Hawthorne, Nathaniel. *The House of Seven Gables.* 1851. Reprint, New York: Bantam, 1981.

Heyrman, Christine Leigh. *Commerce and Culture: The Maritime Communities of Colonial Massachusetts, 1690–1750.* New York: W. W. Norton, 1984.

Hickey, Donald R. *The War of 1812: A Forgotten Conflict.* Urbana: University of Illinois Press, 1989.

Hill, Hamilton Andrew. *Boston's Trade and Commerce for Forty Years, 1844–1884.* Boston: T. R. Marvin, 1884.

Holt, Michael. *The Rise and Fall of the American Whig Party: Jacksonian Politics and the Onset of the Civil War.* New York: Oxford UP, 1999.

———. *By One Vote: The Disputed Presidential Election of 1876.* Lawrence: University of Kansas Press, 2008.

Howe, David Walker. *What Hath God Wrought: The Transformation of America, 1815–1848.* New York: Oxford UP, 2007.

Johnson, Allen, and Dumas Malone. *Dictionary of American Biography*, Vols. 5 and 6. New York: Charles Scribner, 1930.

Johnson, Arthur, and Barry E. Supple. *Boston's Capitalists and Western Railroads: A Study in Nineteenth-Century Railroad Investment Process.* Cambridge, MA: Harvard UP, 1969.

Jorgensen, Neil. *A Guide to New England's Landscape.* Chester, CT: Pequot, 1977.

Kammen, Michael. *Mystic Chords of Memory: The Transformation of Tradition in American Culture.* New York: Vintage, 1993.

Karwatka, Dennis. "The Hoosac Tunnel," in *Tech Directions* 67.9 (April 2008): 3–12.

Kasson, John F. *Civilizing the Machine: Technology and Republican Values in America 1776–1900*. New York: Penguin, 1976.

Kelly, Jack. *Heaven's Ditch: God, Gold, and Murder on the Erie Canal*. New York: St. Martin's, 2016.

Kennedy, Charles J. "The Early Business History of Four Massachusetts Railroads—III," in *Bulletin of Business Historical Society* 25.3 (Sept. 1951): 188–203.

Kirkland, Edward Chase. "The 'Railroad Scheme' in Massachusetts," in *Journal of Economic History*, 5.2 (Nov. 1945): 145–171.

———. "The Hoosac Tunnel: The Great Bore," in *New England Quarterly* 20 (March–Dec. 1947): 88–113.

———. *Men, Cities, and Transportation: A Study in New England History 1820–1900*, Vols. 1 and 2. Cambridge, MA: Harvard UP, 1948.

Knight, Peter. *Reading the Market: Genres of Financial Capitalism in Gilded Age America*. Baltimore: Johns Hopkins UP, 2016.

Labaree, Benjamin W. "The Making of an Empire: Boston and Essex County, 1790–1850," in Conrad Edick Wright and Katheryn P. Viens, eds., *Entrepreneurs: The Boston Business Community, 1700–1850*: 343–363. Boston: Massachusetts Historical Society, 1997.

Lears, Jackson. *Rebirth of a Nation: The Making of Modern America, 1877–1920*. New York: Harper Perennial, 2009.

Lepler, Jessica M. *The Many Panics of 1837: People, Politics, and the Creation of a Transatlantic Financial Crisis*. New York: Cambridge UP, 2013.

Levine, Bruce. "Conservatism, Nativism, and Slavery: Thomas R. Whitney and the Origins of the Know-Nothing Party," in *Journal of American History* 88.2 (Sept. 2001): 402–432.

Levine, Harry. *The Power of Darkness: Hawthorne, Poe, Melville*. Athens: Ohio UP, 1958.

Lewis, Perry. *Boats against the Current: American Culture between Revolution and Modernity, 1820–1860*. Lanham, MD: Rowman & Little, 1993.

Leyda, Jay. *The Melville Log: A Documentary Life of Herman Melville, 1819–1891*. New York: Harcourt & Brace, 1951.

Lord, Francis A. *Lincoln's Railroad Man: Herman Haupt*. Rutherford, NJ: Fairleigh-Dickinson UP, 1969.

Marx, Leo. *The Machine in the Garden: Technology and the Pastoral Ideal in America*. New York: Oxford UP, 1964.

Matthews, Jean V. *Toward a New Society: American Thought and Culture, 1800–1830*. Boston: Twayne, 1991.

McCullough, David. *The Great Bridge*. New York: Simon & Schuster, 1972.

McKey, Richard H., Jr. "Elias Hasket Derby: Merchant of Salem, Massachusetts, 1739–1799." PhD dissertation, Clark University, 1951.

McPherson, James M. *Battle Cry of Freedom: The Civil War Era.* New York: Oxford UP, 1988.

Melville, Herman. *The Confidence-Man.* 1857. Reprint, New York: Signet, 1954.

———. *The Writings of Herman Melville,* Vol. 14. Evanston, IL: Northwestern UP, 1993.

———. *Moby-Dick; or, The Whale.* 1851. Reprint, New York: Putnam, 1998.

Menand, Louis. *The Metaphysical Club: A Story of Ideas in America.* New York: Farrar, Straus & Giroux, 2001.

Merry, Robert W. *A Country of Vast Design: James K. Polk, the Mexican War, and the Conquest of the American Continent.* New York: Simon & Schuster, 2009.

Meyer, William B. "The Long Agony of the Great Bore," in *Invention and Technology* 1.2 (Fall 1985): 18–32.

Middleton, William D. "Hoosac Tunnel: Go Big, Go Deep," in *Trains* 68.11 (Nov. 2008): 14–21.

Miller, Perry. *The New England Mind: The Seventeenth Century.* Boston: Houghton Mifflin, 1954.

Morgan, Edmund S. *The Puritan Dilemma: The Story of John Winthrop.* New York: Longman, 2007.

Morrison, Samuel Eliot. *The Maritime History of Massachusetts, 1783–1860.* Boston: Houghton Mifflin, 1921.

Mulkern, John R. *The Know-Nothing Party in Massachusetts: The Rise and Fall of a People's Movement.* Boston: Northeastern UP, 1990.

North, Douglas C. *The Economic Growth of the United States, 1790–1860.* New York: W. W. Norton, 1966.

O'Connor, Thomas H. *Lords of the Loom: The Cotton Whigs and the Coming of the Civil War.* New York: Scribner's Sons, 1968.

Overton, Richard C. *Burlington Route: A History of the Burlington Lines.* Lincoln: University of Nebraska Press, 1965.

Peterson, Mark. *The City-State of Boston: The Rise and Fall of an Atlantic Power, 1630–1865.* Princeton, NJ: Princeton UP, 2019.

Porter, Glenn. *The Rise of Big Business, 1860–1920.* Wheeling, IL: Davidson, Harlan, 1973.

Potter, David S. *The Impending Crisis, 1848–1861.* New York: Harpers & Row, 1976.

Prude, Jonathon. *The Coming of the Industrial Order: Towns and Factory Life in Rural Massachusetts, 1810–1860.* New York: Cambridge UP, 1983.

Reynolds, David S. *Waking Giant: America in the Age of Jackson.* New York: Harper Collins, 2008.

Rezneck, Samuel. *Business Depressions and Financial Panics: Essays in American Business and Economic History.* Westport, CT: Greenwood, 1971.

Salsbury, Stephen. *The State, the Investor, and the Railroad: The Boston & Albany 1825–1867.* Cambridge, MA: Harvard UP, 1967.

Sankovich, Nina. *The Lowells of Massachusetts: An American Family.* New York: St. Martin's, 2017.

Schexnayder, Cliff. *Builders of the Hoosac Tunnel: Baldwin, Crocker, Haupt, Doane, Shanly.* Portsmouth, NH: Peter E. Randall, 2015.

Scott, James C. *Seeing Like the State: How Certain Schemes to Improve the Human Condition Have Failed.* New Haven, CT: Yale UP, 1998.

Sellers, Charles. *The Market Revolution: Jacksonian America, 1815–1846.* New York: Oxford UP, 1991.

Sheriff, Carol. *The Artificial River: The Erie Canal and the Paradox of Progress, 1817–1862.* New York: Hill & Wang, 1996.

Smith, Henry Nash. *Virgin Land: The American West as Symbol and Myth.* Cambridge, MA: Harvard UP, 1950.

Stiles, T. J. *The First Tycoon: The Epic Life of Cornelius Vanderbilt.* New York: Vintage, 2017.

Stover, John F. *American Railroads.* Chicago: University of Chicago Press, 1961.

———. *History of the Baltimore and Ohio Railroad.* West Lafayette: Univ. of Indiana Press, 1987.

Supple, Barry E. *Boston Capitalists and Western Railroads: A Study in Nineteenth-Century Railroad Investment Process.* Cambridge, MA: Harvard UP, 1969.

Taylor, Alan. *American Colonies: The Settlement of North America.* New York: Penguin, 2001.

Taylor, George Rogers. *The Transportation Revolution, 1815–1860.* New York: Harpers & Row, 1951.

Termin, Peter. *Jacksonian Economy.* New York: W. W. Norton, 1969.

Tyler, John W. "Persistence and Change in Boston Business Community, 1775–1790," in Conrad Edick Wright and Katheryn P. Viens, eds., *Entrepreneurs: The Boston Business Community, 1700–1850.* Boston: Massachusetts Historical Society, 1997: 97–119.

Vickers, Daniel, and Vince Walsh. *Young Men and the Sea: Yankee Seafarers in the Age of Sail.* New Haven, CT: Yale UP, 2005.

Walker, Timothy. "Defense of Mechanical Philosophy," in *North American Review* 33.2 (July 1831): 122–136.

Ward, James A. *That Man Haupt: A Biography of Herman Haupt.* Baton Rouge: Louisiana State UP, 1973.

———. *Railroads and the Character of America, 1820–1887.* Knoxville: University of Tennessee Press, 1986.

Wertenbaker, Thomas Jefferson. *The Puritan Oligarchy: The Founding of American Civilization.* New York: Scribner, 1947.

Wheelwright, William Bond. *The Life and Times of Alvah Crocker.* Boston: Walton, 1923.

White, Richard. *Railroaded: The Transcontinentals and the Making of Modern America.* New York: W. W. Norton, 2011.

———. *The Republic for Which It Stands: The United States during Reconstruction and the Gilded Age, 1865–1896.* New York: Oxford UP, 2017.

Wilentz, Sean. *The Rise of American Democracy: Jefferson to Lincoln.* New York: W. W. Norton, 2005.

Wolmar, Christian. *The Great Railroad Revolution: The History of Trains in America.* New York: Public Affairs, 2012.

WPA Guide to Massachusetts, The. 1937. Reprint, New York: Pantheon Books, 1983.

Young, Alfred F. *The Shoemaker and the Tea Party: Memory and the American Revolution.* Boston: Beacon Press, 1999.

INDEX

Fowle, Joseph, 103, 123, 124, 125–26
Fremont, John, 96
Frost, Benjamin, 150–51
fugitive slaves, 56, 92, 99, 113, 197n16
Fulton, Robert, 32

Galbraith, William, 87–88, 89, 103
Gallatin, Albert, 27
Gardner, Henry J., 91, 94, 95–96
Garrison, William Lloyd, 113
gasometer, 136, 137
Gaston, William, 169, 174
glaciation, 7–8
Goodwin, George, 137
Gore, Christopher, 30
Gould, Jay, 150, 162, 173
Graham, Sylvester, 91
Grand Trunk Railroad, 145, 186, 212–13n4
Granger, W. P., 133, 138, 167, 174, 205n27
Granger Laws, 212n43
Grant, Anne, 23
Grant, Ulysses S., 144, 149, 163, 207n24
Great Debate of 1873, 160–62
Great Fire of Boston, 155, 168
Great Migration, 9–10, 12
Great Strike of 1877, 179
Greenfield, MA, 57, 106
Green River Bridge collapse, 106, 108, 109, 148
Griswold, Whiting, 53–54, 162
Gwynn, Stuart, 103, 124, 125

Hale, Nathan, 34–35, 40–41
Hamilton, Edward (Theseus), 140–41, 206n10
Hancock, John, 30
Harper's Ferry raid, 113, 201n31
Harris, Daniel, 90, 95, 99, 107
Harsen, William, 124, 125–26
Haskins, Edwin, 137
Haupt, Herman: accusations against, 90–91, 99, 104, 109, 111; biographical background, 87–88; Bird attacks on, 108–10; Bird's reconciliation with, 202n39; and Crocker, 108, 114; and drilling machines, 103, 201n28; fight for political support by, 97–98, 107–8, 200n20; financing sought by, 89–90, 93, 94–95, 97, 98–100, 105–6; and Hoosac Tunnel construction, 102, 116; later life of, 111, 202n42; legacy of, 111–12; and legislative committee fight, 107–8; and Lincoln, 5, 108; management style and weaknesses of, 104, 112, 119; and mechanical drill, 124–25, 204n18; military career of,

110–11; no Hoosac Tunnel without, 112, 187; origin of involvement with Hoosac Tunnel, 87–89; perseverance of, 97, 187; personal investment in tunnel project by, 89, 90, 96; personality of, 112; photos of, 75, 87; salesmanship of, 97–98, 100; and Serrell, 198n30, 201n28; Troy & Greenfield stock held by, 98–99, 211–12n40
Haupt, Jacob, 124, 125
Hawthorne, Nathaniel, 195n38; *The House of Seven Gables*, 42–44, 162
Hayes, Rutherford, 211n36
Hayward, James, 59
Higginson, Thomas Wentworth, 113
Hitchcock, Edward, 5, 69; geological analysis by, 51, 52, 54, 63, 166, 184, 196n8
Hocking, James, 147
Hofstadter, Richard, 199n12
Holmes, Alexander, 114–15, 130
Hooker, Richard, 10
Hoosac line, 178, 181, 182, 186
Hoosac Mountain, 2, 58, 204n14; formation of, 101; geological composition of, 51, 52, 63, 166; idea of tunnel-canal through, 31, 52, 93–94, 194n20; Pocumtucks' naming of, 57
Hoosac Tunnel: Andrew commission on, 110, 111, 114–17; Bird pamphlets attacking, 108–9, 116, 127–28, 138–39; and Civil War, 113–14; Crocker as "father" of, 5, 45, 50–51; Crocker as lobbyist for, 128–29, 139; first trains through, 174; Fitchburg Railroad purchase of, 180–81; forgotten in history, 3, 188; as "Great Bore," 5, 74; Haupt political fight for, 97–98, 99–100, 105–6, 107–8, 200n20; Hitchcock geological review of, 51, 52, 54, 63, 166, 184, 196n8; idea first raised of, 50–51; investors in, 51–52, 62, 87–88, 89–90, 97, 98, 103–4; maintenance and upkeep of, 177, 180; misjudgments about, 172–73, 184, 186–87; newspaper attacks on, 52–53, 63–64, 128; popular support for, 129; Port of Boston revitalized by, 185–86; and public opinion, 51–52, 97–98, 107–8, 129, 200n20; railroad use of completed, 179–80; size of, 116, 211n33; state financial aid to, 51, 55, 56–62, 91, 93–95, 98–99, 100, 107, 108, 115, 117, 128, 140, 141; today, 183, 188; West's lure as driving forcer, 60–61, 169–70. *See also* Massachusetts legislature debates
Hoosac Tunnel construction: accidents and deaths in, 3, 101–2, 106, 120–21, 127, 136, 137–38, 156, 187, 190n8; alignment of, 116, 119, 122–23; arching in, 51, 78, 116, 133, 158, 159, 165–68, 180, 184; brick kiln for, 83, 117, 132; brick tube in, 78, 132, 138,